rings, extensions, and
cohomology

PURE AND APPLIED MATHEMATICS

A Program of Monographs, Textbooks, and Lecture Notes

LECTURE NOTES IN PURE AND APPLIED MATHEMATICS

Additional Volumes in Preparation

rings, extensions, and cohomology

**proceedings of the conference on the occasion
of the retirement of Daniel Zelinsky**

edited by
Andy R. Magid
The University of Oklahoma
Norman, Oklahoma

CRC Press
Taylor & Francis Group
Boca Raton London New York

CRC Press is an imprint of the
Taylor & Francis Group, an **informa** business

CRC Press
Taylor & Francis Group
6000 Broken Sound Parkway NW, Suite 300
Boca Raton, FL 33487-2742

First issued in hardback 2017

© 1994 by Taylor & Francis Group, LLC
CRC Press is an imprint of Taylor & Francis Group, an Informa business

No claim to original U.S. Government works

ISBN 13: 978-1-138-40205-8 (hbk)
ISBN 13: 978-0-8247-9241-1 (pbk)

Visit the Taylor & Francis Web site at
http://www.taylorandfrancis.com

and the CRC Press Web site at
http://www.crcpress.com

Library of Congress Cataloging-in-Publication Data

Rings, extensions, and cohomology : proceedings of the conference on the occasion of the retirement of Daniel Zelinsky / edited by Andy R. Magid.
    p. cm. — (Lecture notes in pure and applied mathematics; v. 159)
    "Papers delivered at a Conference on Rings, Extensions, and Cohomology, held at Northwestern University, August 29 and 30, 1993"—Galley.
    Includes bibliographical references and index.
    ISBN 0-8247-9241-6
    1. Ring extensions (Algebra)—Congresses. 2. Galois theory—Congresses. 3. Homology theory—Congresses. I. Zelinsky, Daniel. II. Magid, Andy R. III. Conference on Rings, Extensions, and Cohomology (1993 : Northwestern University, Evanston, Ill.) IV. Series.
QA247.R5752 1994
512'.4—dc20

94-11105
CIP

Preface

The papers in this volume were delivered at a conference on Rings, Extensions, and Cohomology, held at Northwestern University, 1993, on the occasion of the retirement of Daniel Zelinsky. In a few cases, the contributors were unable to attend the conference in person and submitted their papers for this volume only.

The subjects covered range across commutative and noncommutative ring theory, especially separability and Galois theory, including Lie algebra and module theory: topics which have been the core of the mathematical interests of Daniel Zelinsky, as well as his students, and their students, who are represented here.

The papers of T. Kambayashi, A. Magid, and A. Simis *et al* deal primarily with commutative ring theory, while those of S. Amitsur, D. Haile, M. Rosset and S. Rosset, and A. Rosenmann are primarily in noncommutative ring theory, and W. Brown's is in both. A. Fauntleroy's paper is on algebraic geometry, extending some commutative algebra methods to sheaves. The papers of L. Childs, H. Kreimer, and T. McKenzie are in the general area of separability and Galois theory for commutative rings, and those of R. Alfaro and G. Szeto, M. Beattie, and S. Ikehata and G. Szeto deal with generalizations of Galois theory. J. Wehlen's paper considers certain noncommutative algebras over commutative rings. L. Kadison's paper studies an important case of noncommutative separability. The papers of J. Bergen and P. Grzeszczuk, R. Dahlberg, and G. Nelson are on aspects of Lie theory and that of M. Goddard on module theory.

Doctoral students of Daniel Zelinsky contributing papers to this volume are William C. Brown, T. Kambayashi, Andy Magid, and Shmuel Rosset. Doctoral students of doctoral students of Daniel Zelinsky contributing papers are Amnon Rosenmann, student of Shmuel Rosset, and Randall Dahlberg and Graydon Nelson, students of Andy Magid.

All the conference attendees noted the vitality of the subject, not only among the veteran scholars who spoke but also among the new and recent PhDs who presented papers, a vitality matched by that of the conference honoree. On behalf of all the contributors, we dedicate this volume to our friend and teacher, Daniel Zelinsky.

Andy Magid

Contents

Contributors

RICARDO ALFARO   University of Michigan-Flint, Flint, Michigan

S. A. AMITSUR   Hebrew University, Jerusalem, Israel

M. BEATTIE   Mount Allison University, Sackville, New Brunswick, Canada

JEFFREY BERGEN   DePaul University, Chicago, Illinois

WILLIAM C. BROWN   Michigan State University, East Lansing, Michigan

LINDSAY N. CHILDS   State University of New York at Albany, Albany, New York

RANDALL P. DAHLBERG   Allegheny College, Meadville, Pennsylvania

AMASSA FAUNTLEROY   North Carolina State University, Raleigh, North Carolina

MARK A. GODDARD   The University of Akron, Akron, Ohio

PIOTR GRZESZCZUK   University of Warsaw, Białystok Division, Białystok, Poland

DARRELL E. HAILE   Indiana University, Bloomington, Indiana

SHÛICHI IKEHATA   Okayama University, Okayama, Japan

LARS KADISON   Roskilde University, Roskilde, Denmark

T. KAMBAYASHI   Tokyo Denki University, Hatoyama, Saitama-ken, Japan

H. F. KREIMER   Florida State University, Tallahassee, Florida

ANDY R. MAGID   University of Oklahoma, Norman, Oklahoma

THOMAS McKENZIE   Bradley University, Peoria, Illinois

GRAYDON NELSON   University of Oklahoma, Norman, Oklahoma

AMNON ROSENMANN   University of Essen, Essen, Germany

MYRIAM ROSSET   Bar-Ilan University, Ramat Gan, Israel

SHMUEL ROSSET   Tel-Aviv University, Ramat-Aviv, Tel-Aviv, Israel

ARON SIMIS   Universidade Federal da Bahia, Salvador, Bahia, Brazil

GEORGE SZETO   Bradley University, Peoria, Illinois

BERND ULRICH   Michigan State University, East Lansing, Michigan

WOLMER V. VASCONCELOS   Rutgers University, New Brunswick, New Jersey

JOSEPH A. WEHLEN   Computer Sciences Corporation, Integrated Systems Division, Moorestown, New Jersey

Daniel Zelinsky: An Appreciation

It has fallen to me, on this occasion of his retirement from Northwestern University after almost four and a half decades of service in the Department of Mathematics, to offer a few comments on Dan Zelinsky: The Scholar and the Teacher, So Far.

Actually, this is going to be a pretty seamless retirement. Dan will be back teaching at Northwestern in the upcoming academic year, as he has every fall since 1949. Nonetheless, retirements are significant milestones, and the obligatory speech at the retirement banquet is a significant responsibility, so before assuming it I will relate a story Dan told me when I was a graduate student.

This involves the great French mathematician Jean Dieudonné, who was on the Northwestern faculty when Dan Zelinsky was a young faculty member. Dieudonné had written a one line review for *Math Reviews* the gist of which was "Lemma 1.1 is wrong and the rest of the paper depends on it". Dan looked up the paper, and found that while indeed Lemma 1.1 was false in the standard terminology, if one used the terminology as the author had introduced it in Definition 1.0, the lemma, and the rest of the paper, was correct, and at least moderately interesting. Zelinsky pointed this out to Dieudonné, who made some mild comment about perhaps writing a replacement review. Zelinsky was worried that this was insufficient to repair the damage to the author's career, to which Dieudonné replied "Look, if the man is a real mathematician, then he'll write other papers, and the review won't matter. And if he's not a real mathematician, it doesn't matter anyway."

I like this story for what it says about Dieudonné, for what it says about Zelinsky, and also for the principle it espouses, which in this context means that at the retirement banquet for a real mathematician, like the present occasion, it doesn't matter whether one gets the speech right or not.

Thus relieved of that reponsibility, I'll begin with *yichus*, which is a Jewish word meaning Ph.D. genealogy. American universities began awarding the doctorate in the last third of the nineteenth century. The mathematics professor at Yale at that time was Hubert Anson Hewton, who supervised the thesis of Eliakim Hastings Moore, who taught briefly at Northwestern before going to the University of Chicago, where he supervised the thesis of Leonard Eugene Dickson, who supervised the Chicago thesis of Abraham Adrian Albert, who supervised the 1946 Chicago thesis of Daniel Zelinsky. At each link in this chain, by the way, the son is one of the most distinguished students of the father; in Dan's case my reference for this is Nathan Jacobson's 1974 *AMS Bulletin* obituary of Albert.

In fact Daniel Zelinsky's professional mathematical career began before his Ph.D. with his wartime service in the AMG–C (Applied Mathematics Group, Columbia). According to the personnel list in Saunders MacLane's historical memoir about this group, Dan was their youngest member.

From an address at a banquet in honor of Daniel Zelinsky, Evanston, Illinois, August 29, 1993

Dan Zelinsky's published work spans four decades and ranges across commutative and noncommutative rings, topological methods in algebra, and cohomology, with Galois theory and Brauer groups being recurring themes. An especially significant chunk is his 10 paper collaboration with Alex Rosenberg from the mid–1950's to the early 1960's, including a paper in the famous Dimension Club series in the *Nagoya Journal* (On the Dimensions of Modules and Algebras VIII), in which they were joined by Samuel Eilenberg. That particular authorial trio – Eilenberg, Rosenberg, and Zelinsky – is not only an euphonious joy to recite, but of course significant contributors to postwar algebra.

Dan's 1953 *American Journal* paper "Linearly compact modules and algebras" is still regularly cited, as I discovered with a brief search under "Zelinsky" in the *Science Citation Index*. In fact it is instructive to look at the index of that 1953 volume, where one finds the names Baer, Buck, Borel, Carlitz, Cassels, Conrad, Dieudonné, Eilenberg, Goldhaber, Hartman, Herstein, Kaplansky, Kolchin, MacLane, Mackey, Rosenberg, Rosenlicht, Thrall, and Wintner, before reaching Zelinsky. One wonders if any current journal will be able to match that for name recognition 40 years from now.

Dan also found time to write a widely used textbook on linear algebra, which allows me to move on to some comments about Daniel Zelinsky as a teacher. It was 50 years ago this fall that Dan began his classroom career as a University of Chicago Mathematics Instructor, which he was again in 1946–47. In fact the Department of Mathematics schedule for that year, a copy of which was retrieved by Paul Sally, reveals that for the autumn quarter Dan Zelinsky taught Math 101c at 8:30 and Math 215c at 9:30 every day except Wednesday, with similar assignments at those days and times for the winter and spring quarters also. After spending the next two academic years at the Institute for Advanced Study, Dan accepted a position at Northwestern University for the academic year 1949–50, and there he has remained ever since. A conservative order of magnitude estimate is that he has taught 300 classes at Northwestern, of which I took one. I'd like to comment on that (possibly unrepresentative) sample.

I refer to second year graduate algebra, Math D32, in academic year 1966-67. Dan's teaching style that year ran to the simple graphic. Here's an example. Suppose we were studying the theorem that a projective module over a local ring is free. Dan would state the theorem in telegraphic style:

Th. *R local w/ M. P free ⇔ proj*

Then he would proceed with the proof. One implication here ($\Rightarrow$) is obvious and would be so noted with implication symbol and a checkmark. The discussion of the other implication ($\Leftarrow$) would begin by entering that symbol in the proof, and then mathematical details would be presented. That particular year, while Dan was doing the details, which were generally all verbal, he would illustrate by sketching to the side on the chalkboard a tight spiral, whose length was dictated by the complexity of the proof. Once achieved, that would be noted by a checkmark after the implication symbol also. Thus finally the chalkboard would display the proof:

Prf.

$$\implies \quad \checkmark$$
$$\impliedby \quad \checkmark$$

Of course immediately after class we students would rush to the library and translate our class notes into a complete mathematical record, and in the process really learn the subject. What I have never learned is whether this was by Dan's design or just happy circumstance.

Part of the year Dan was following texts for the course (Van der Waerden, Zariski and Samuel), and part of the time not. In the latter phases, he had the students take turns making and distributing class notes, which he reviewed before circulating. I was assigned the separable algebra section, a subject about which I thought considerably in later years. At this time, which was my first exposure, I was just eager to do my assignment well, so I was happy when Dan returned my notes to me for modification before circulation to see he'd put a big "A" in red pencil on the top; I hadn't even known they were to be graded. In fact they weren't: the "A" was in my title, which I had misspelled "Scperable Algebra", an error I repeated and Dan corrected 20 or so times in that document (which I still possess), and have been subsequently able to avoid ever since.

Dan Zelinsky accepted me as a doctoral student at the end of that year, and the following December I took the oral exams for that program. My examining committee consisted of Leonard Evens, Eben Matlis, and Daniel Zelinsky. One of my topics was semisimplicial complexes, and I recall that at one point in the midst of some complicated explanation of that subject I turned from the blackboard to see that all three of my examiners were slumped down in their chairs chewing on the ends of their neckties. Naturally, that scene made an indelible impression on me, but when I conjure it up now, what makes it clear that 26 years have passed is that all three examiners were *wearing* neckties.

I passed that exam, and began my thesis work. Like all Dan's students, I regularly reported to his corner office on the second floor of Lunt Hall, where I stood at the blackboard with views out the windows of the library lawn and Sheridan Road, and explained my latest results while Dan lay on his office couch and assumed his working posture. (Dan's son Paul has explained how when he was a boy he would come home to see his father stretched out on the sofa with his eyes closed and be warned by his mother to "be quiet, your father is working.")

I remember Dan's characterization of one of my earliest attempts at a theorem, which he described as "If we had some ham, we could have some ham and eggs, if we had some eggs".

Of course I finished and graduated, one of 14 Ph.D. students Dan Zelinsky supervised over 30 years. During his years at Northwestern he also served on the Council of the American Mathematican Society and edited its *Transactions*, as well as chairing the Mathematics Section of the American Association for the Advancement of Science, and the Northwestern Mathematics Department, and performed a long list of other services to his profession and university.

One treat about addressing a general audience is the opportunity to tell old math jokes. For example, the one about the three kinds of mathematicians: those who can count and those who can't. [*Note: the audience laughed here; whether out of novelty or politeness is not known.*]

Then there is the kind who really count, like Dan Zelinsky. We read in the section of the *Talmud* called *Avot* the instruction to "find yourself a teacher, get yourself a friend to study with". I know I speak for Dan's students and colleagues in saying that in Daniel Zelinsky we got both.

Andy R. Magid

rings, extensions, and
cohomology

The Centralizer on H-Separable Skew Group Rings

RICARDO ALFARO, Mathematics Department, University of Michigan-Flint, Flint, MI 48502.

GEORGE SZETO, Mathematics Department, Bradley University, Peoria, Il 61625.

1 INTRODUCTION

A ring A is said to be *H-separable* over a subring B if $A \otimes_B A$ is isomorphic to a direct summand of n copies of A as $(A\text{-}A)$-bimodules for some positive n. It follows that Δ, the centralizer of B in A, is a finitely generated projective C-module, where C is the center of A; and also that $A \otimes_B A$ is isomorphic to $\mathrm{Hom}_C(\Delta, A)$ as $(A\text{-}A)$-bimodules. In fact, these last two conditions together are equivalent to A being H-separable over B.

A good deal of attention has been paid to the H-separability condition for simple, primitive and skew polynomial rings in [Sug82], [Sug87], [Sug90], [Ike90], [Ike91], and others. In [Alf92] we started the study of H-separability conditions of the skew group ring $S*G$ over the base ring S. In the case S is a commutative ring it is known that $S*G$ is H-separable over S if and only if S is a G-Galois extension of S^G. The aim of this paper is to continue [Alf92] and establish the Galois condition on the centralizer of the ring S in $S*G$ for non-commutative rings with finite group actions. The center of a ring R will be denoted by $Z(R)$, and all rings are assumed to have a unity.

This project was supported in part by a grant from the Faculty Research Initiative Program of the University of Michigan-Flint.

2 THE GALOIS CONDITION

We say that a finite group G *acts faithfully* on a ring S if there is a group monomorphism from G to $\text{Aut}(S)$. We will denote by ${}^g r$ the image of $r \in S$ under $g \in G$. The fixed ring S^G is the set of all elements of S fixed by every element of G. We say $a_1, \ldots, a_n; b_1, \ldots, b_n$ (elements in S) is a *G-Galois basis* for S if $\sum_i a_i\, {}^g b_i = \delta_{1,g}$ for all $g \in G$. (Here $\delta_{i,g}$ is the Kronecker delta.) For notation, we will write $\{a_i, b_i\}$ is a G-Galois basis, and when such a set exists, we say that S is G-Galois over S^G.

For any ring S and a finite group G acting faithfully on S, the *skew group ring* is an associative ring $S*G$ with identity, where every element $\alpha \in S*G$ can be written uniquely as a sum $\alpha = \sum_{g \in G} s_g g$ with $s_g \in S$. The addition is obvious and multiplication is given by the formulas $gs = {}^g s g$ for all $s \in S, g \in G$. Consider S as a right S^G-module, and as a left $S*G$-module with the action of $S*G$ on S given by $(sg) \cdot r = s\, {}^g r$. The ring S is G-Galois over S^G if and only if S is finitely generated projective as a right S^G-module and the map $j : S*G \to \text{End}(S_R)$ induced by the action of $S*G$ on S is a ring isomorphism.

Given a ring S and the skew group ring $S*G$, the centralizer of S in $S*G$, denoted by Δ, is the set of all elements in $S*G$ which commute with every element of S. The group G induces an action on the skew group ring by conjugation; furthermore since Δ is G-invariant, it induces an action on Δ. It is easy to see that the fixed ring Δ^G is the center of $S*G$, denoted by C. The first relation between the H-separability of the skew group ring and the centralizer is given by the following theorem, which has appeared in [Alf], but we give here a different proof using H-systems as defined in [NS75]. The extension A over B is H-separable if there exist some $v_i \in C_A(B)$ and $d_i \in C_{A \otimes_B A}(A)$, called an H-system, such that $\sum_i v_i d_i = 1 \otimes 1$.

THEOREM 1. *If Δ is G-Galois over C, then $S*G$ is H-separable over S.*

Proof. Let $\{a_i, b_i\}$ be a G-Galois basis for Δ over C. Let $x_{ig} = gb_i$ and $y_{ig} = g^{-1}$ elements of $S*G$. We claim that $\{a_i, \sum_g x_{ig} \otimes y_{ig}\}$ is an H-system and so $S*G$ is H-separable over S.

First we need to show that $\sum_g x_{ig} \otimes y_{ig}$ is in the centralizer of S in $S*G \otimes_S S*G$:

$$sh\left(\sum_{g \in G} x_{ig} \otimes y_{ig} \right) = \sum_{k \in G} skb_i \otimes k^{-1}h = \sum_{k \in G} kb_i \otimes {}^{k^{-1}} s k^{-1} h = \sum_{k \in G} kb_i \otimes k^{-1} sh = \left(\sum_{g \in G} x_{ig} \otimes y_{ig} \right) sh.$$

But also, $\sum_{i,g} a_i x_{ig} \otimes y_{ig} = \sum_{i,g} a_i\, {}^g b_i g \otimes g^{-1} = \sum_g \delta_{1,g} g \otimes g^{-1} = 1 \otimes 1$, hence we obtained an H-system for $S*G$ over S. $\square$

In most of the results about H-separable extensions A of a ring B, it has been shown that the double centralizer property plays a very important role; and it is no surprise that the main results of this work depend on the centralizer of the skew group ring satisfying a double centralizer property. In preparation for it, we now give a general proposition for group actions that must be well known but we couldn't find in the literature.

PROPOSITION 1. _Let G be a finite group acting faithfully on a ring R. Assume the action is G-Galois and assume R^G is a commutative ring; then R satisfies the double centralizer property in $R*G$, (i.e., $C_{R*G}(C_{R*G}(R)) = R$), and hence $Z(R*G) = Z(R)^G$._

Proof. Since R^G is commutative and the action is G-Galois , then R is an R^G- progenerator. Hence $\text{Hom}_{R^G}(R, R)$, and so $R*G$, is an Azumaya algebra over R^G. But also R is separable over R^G (see [Alf]), and hence by [OKI87], R satisfies the double centralizer property. Furthermore, if $T = C_{R*G}(R)$ then we have $Z(R*G) = T^G = Z(T) \cap T^G = Z(T)^G = (C_{R*G}(T) \cap T)^G = (R \cap T)^G = Z(R)^G$. $\square$

The group action of G on S induces an action on the centralizer Δ by conjugation. Although they are different automorphism groups, we can still say that G acts on Δ, and we will use the same notation for both groups unless it is neccesary to do otherwise. When this action is faithfull, we can form the skew group ring $\Delta*G$. Since the fixed ring Δ^G coincides with the center C of $S*G$, proposition 1 gives the following corollary.

COROLLARY 1. _If Δ is G-Galois over C, then Δ satisfies the double centralizer property in $\Delta*G$, and $C = Z(S)^G$._

The next result is a direct consequence of a lemma by Ikehata [Ike81], which indicates that every unital left A-module which is a generator as a left B-module, is also a generator as left A-module (where A is an H-separable extension of B.)

PROPOSITION 2. _If the skew group ring $S*G$ is H-separable over S, then S is G-Galois over S^G._

Proof. By [Ike81], S is a generator as a left $S*G$-module. Thus by Morita theorem, $S*G \cong \text{End}_{S^G}(S)$ and S is a finitely generated projective S^G-module. Hence the action is G-Galois on S. $\square$

Using this proposition, we can now show that for the centralizer Δ the Galois condition is equivalent to the H-separability condition of the corresponding skew group ring. This result is the analogue in the non-commutative case to the well known fact that if the ring S is commutative, the skew group ring is H-separable over S if and only if S is G-Galois over S^G.

THEOREM 2. _The centralizer Δ is G-Galois over C if and only if $\Delta*G$ is H-separable over Δ._

Proof. As shown in the proof of proposition 1, $\Delta*G$ is an Azumaya C-algebra; but $\Delta*G$ is free as a left Δ-module, hence $\Delta*G$ is H-separable over Δ by [Ike81]. The converse follows now from proposition 2. $\square$

3 THE H-SEPARABLE CONDITION

We now study under which conditions the H-separability of the skew group ring $S*G$ over S implies that the commutator Δ is a G-Galois extension. The fact that the commutator is G-Galois has been used to prove a Noether-Skolem type theorem for group-graded rings by Osterburg and Quinn in [OQ88]. All the notation from the previous section is assumed.

PROPOSITION 3. *Let $S*G$ be H-separable over S. If Δ satisfies the double centralizer property in $\Delta*G$, then Δ is G-Galois over C.*

Proof. We first show that G acts faithfully on Δ. The ring S being a direct summand of $S*G$ as left S-modules, satisfies the double centralizer property on $S*G$ by [Sug67]. Thus, if $g \in G$ and ${}^g d = d$ for all $d \in \Delta$, then $gd = dg$ and so $g \in C_{S*G}(\Delta) = S$, forcing $g = 1$.

Since $S*G$ is separable over S, there exists a central element in S of trace one [Alf]. Since S satisfies the double centralizer property in $S*G$, then $Z(\Delta) = Z(S)$, and $\Delta^G = Z(S)^G$; but also $C_{\Delta*G}(C_{\Delta*G}(\Delta)) = \Delta$, thus $Z(\Delta*G) = Z(\Delta)^G$. Hence there is a central element in Δ of trace one, so $\Delta*G$ is separable over Δ, thus also over $\Delta^G (= Z(S)^G = Z(\Delta*G))$. Therefore $\Delta*G$ is an Azumaya Δ^G-algebra. Furthermore Δ is a direct summand of $\Delta*G$ as $(\Delta-\Delta)$-bimodules, thus Δ is separable over Δ^G and $\Delta*G$ is H-separable over Δ by [OKI87]. Proposition 2 now says that Δ is G-Galois over $\Delta^G (= C.)$ □

Now we put together this last propposition with the results from the previous section, and we can prove the main result about equivalent conditions for the centralizer of the skew group ring to be a G-Galois extension. This theorem resembles the result in [Ike81] for group actions over commutative rings.

THEOREM 3. *With all the notation as above; the following are equivalent:*
  (1) Δ *is G-Galois over C.*
  (2) $\Delta*G$ *is H-separable over Δ.*
  (3) $S*G$ *is H-separable over S, and Δ satisfies the double centralizer property on $\Delta*G$.*
  (4) $\Delta*G$ *is an Azumaya C-algebra.*

Proof. The equivalence of the first two statements is theorem 2. The equivalence of the frist and third statement is theorem 1, corollary 1 and proposition 3. In the proof of proposition 3 we actually proved that (3) implies $\Delta*G$ is an Azumaya C-algebra and this implies Δ is G-Galois ; thus the equivalence follows immediately.   □

Example. Let S be the division ring of quaternions, $S = \mathbb{R} \cdot 1 + \mathbb{R} \cdot i + \mathbb{R} \cdot j + \mathbb{R} \cdot k$. The group $G = \{1, i, j, k\}$ acts on S by conjugation. To differentiate between the element of G and the element of S we write $\hat{i} \in G$, and so on. It is easily seen that the fixed ring is $\mathbb{R}$, and the center of $S*G$ is also $\mathbb{R}$. By [Ike81] the skew group ring $S*G$ is H-separable over S. The centralizer of S in $S*G$ is: $\Delta = \mathbb{R} \cdot \hat{1} + \mathbb{R}i \cdot \hat{i} + \mathbb{R}j \cdot \hat{j} + \mathbb{R}k \cdot \hat{k}$. The group G also acts in Δ by conjugation, and $\Delta^G = \mathbb{R}$. As the center of $\Delta*G$ is also $\mathbb{R}$ and there is an element of Δ with trace 1, then $\Delta*G$ is H-separable over Δ. Hence, by the theorem we know that Δ is G-Galois over $\mathbb{R}$. In fact, we obtain a Galois basis given by

$\{a_1 = 1/2,\ a_2 = (1/2)i\hat{i},\ a_3 = (1/2)j\hat{j},\ a_4 = (1/2)k\hat{k};\ b_1 = 1/2,\ b_2 = -(1/2)i\hat{i},\ b_3 = -(1/2)j\hat{j},\ b_4 = -(1/2)k\hat{k}\}.$

When the double centralizer property of Δ cannot be checked directly, it is still possible to obtain the *G*-Galois condition on the commutator Δ if there is a particular *H*-system for $S * G$ over S.

THEOREM 4. *If $S * G$ is H-separable over S, and there exists an H-system $\{d_i, w_i\}$ with $d_i \in \Delta, w_i = \sum_{g \in G} ga_i \otimes g^{-1}$, then Δ is G-Galois over C.*

For the proof of the theorem we need to show first some conditions satisfied by the elements of $C_{S*G \otimes_S S*G}(S * G)$,:

LEMMA 1. *If $\sum\limits_{\substack{g \in G \\ h \in G}} \alpha_{(g,h)} g \otimes h \ \in \ C_{S*G \otimes_S S*G}(S * G)$, then:*

(1) $\alpha_{(g,h^{-1})} g h^{-1} \in \Delta$ *for all $g, h \in G$.*
(2) $\alpha_{(g,g^{-1})} = {}^g\alpha_{(1,1)}$ *for all $g \in G$.*

Proof of lemma.. Since $k \in G$, commutes with the elements in $C_{S*G \otimes_S S*G}(S * G)$, then:

$$\sum_{\substack{g \in G \\ h \in G}} {}^k\alpha_{(g,h)} kg \otimes h = \sum_{\substack{g \in G \\ h \in G}} \alpha_{(g,h)} g \otimes hk.$$

Making a change of indices we obtain:

$$\sum_{\substack{g \in G \\ h \in G}} {}^k\alpha_{(k^{-1}g,h)} g \otimes h = \sum_{\substack{g \in G \\ h \in G}} \alpha_{(g,hk^{-1})} g \otimes h.$$

Hence ${}^k\alpha_{(k^{-1}g,h)} = \alpha_{(g,hk^{-1})}$ for all $g, h, k \in G$.   (1)
In particular, for $h = 1, g = k$, we obtain $\alpha_{(g,g^{-1})} = {}^g\alpha_{(1,1)}$.
On the other hand, every element $x \in S$ also must commute with the elements in $C_{S*G \otimes_S S*G}(S * G)$, thus:

$$\sum_{g \in G} x\alpha_{(g,1)} g \otimes 1 = \sum_{g \in G} \alpha_{(g,1)} g \otimes x = \sum_{g \in G} \alpha_{(g,1)} {}^g x g \otimes 1.$$

Therefore, $x\alpha_{(g,1)} = \alpha_{(g,1)} {}^g x$ for all $g \in G, x \in S$, and hence $\alpha_{(g,1)} g \in C_{S*G}(S) = \Delta$, for all $g \in G$.
Thus $\alpha_{(k^{-1}g,1)} k^{-1} g \in \Delta$, but Δ is *G*-invariant, so applying k to the last expression and applying (1), we have:

$${}^k(\alpha_{(k^{-1}g,1)} k^{-1} g) = \alpha_{(g,k^{-1})} k k^{-1} g k^{-1} = \alpha_{(g,k^{-1})} g k^{-1} \ \in \Delta \ \text{ for all } g, k \in G.$$

$\square$

Proof of theorem 4. By lemma 1, ${}^g a_i \in \Delta$, thus $a_i \in \Delta$ since Δ is *G*-invariant. Hence $a_i \in Z(S)$. On the other hand, $\sum_{i,g} d_i ga_i \otimes g^{-1} = 1 \otimes 1$. Thus, $\sum_i d^g a_i = \begin{cases} 1 & \text{if } g = 1, \\ 0 & \text{if } g \neq 1. \end{cases}$ Where $d_i \in \Delta$, and $a_i \in Z(S) \subseteq \Delta$. This makes $\{d_i, a_i\}$ a *G*-Galois basis for Δ over C. $\square$

In the general situation, when $S*G$ is H-separable over the ring S, we have seen that S satisfies the double centralizer property and this implies that $Z(S) = Z(\Delta)$, which we will denote by Z; and $Z(S*G) = \Delta^G = Z(S)^G$, which we denote C. Hence we have the following ring inclusions: $C \subseteq Z \subseteq \Delta$. We analize now what happens when each of these inclusions are equalities.

CASE 1: $Z = \Delta$.

An element $g \in G$ is said to be *w-inner* if $\phi_g = \{s \in S/s\, {}^gx = xs \text{ for all } x \in S \} \neq \emptyset$; and G is *w-outer* if the only w-inner element is the identity. It is clear that G is w-outer if and only if $\Delta = Z(S)$ (hence Δ is commutative). In [Alf] it was shown that in this case the commutator Δ is G-Galois over C if and only if $S*G$ is H-separable over S, and thus the ring S is a central Galois extension. We give now an alternative proof of this result using theorem 4, because in this case we obtain such a peculiar H-system.

PROPOSITION 4. *Let $S*G$ be H-separable over S, and assume G is w-outer (equivalently Δ is commutative), then $\Delta = Z(S)$ is a G-Galois extension of C.*

Proof. By lemma 1, if $x = \sum_{\substack{g \in G \\ h \in G}} \alpha_{(g,h)}g \otimes h \in C_{S*G \otimes_S S*G}(S*G)$, then $\alpha_{(g,h)}gh \in \Delta = Z(S)$. Hence $\alpha_{(g,h)} = 0$ if $gh \neq 1$, and therefore

$$x = \sum_{g \in G} \alpha_{(g,g^{-1})}g \otimes g^{-1} = \sum_{g \in G} {}^g\alpha_{(1,1)}g \otimes g^{-1} = \sum_{g \in G} g\alpha_{(1,1)} \otimes g^{-1}.$$

Thus, by theorem 4, Δ and hence $Z(S)$ is G-Galois over $Z(S)^G$. $\square$

CASE 2: $C = Z$.

In this case we don't always get the commutator Δ to be G-Galois , but we show that it is an Azumaya extension; furthermore, the converse also holds.

THEOREM 5. *The skew group ring $S*G$ is H-separable over S and G acts trivially on the center of S if and only if Δ is an Azumaya C-algebra and $S*G = S \cdot \Delta$.*

Proof. (Sufficiency.) Since S is a direct summand of $S*G$ as $(S$-$S)$-bimodules, Δ is a separable C-algebra by corollary 4.2 in [Hir69]. But G acts trivially on $Z(S)$, hence $C = Z(S)^G = Z(S) = Z(\Delta)$, thus Δ is Azumaya. Furthermore, considering $S*G$ as a $(\Delta$-$\Delta)$-bimodule, $\Delta \otimes_C C_{S*G}(\Delta) \cong S*G$, by theorem 1.2 in [Sug67]; but, as indicated in the proof of proposition 3, $C_{S*G}(\Delta) = S$; hence $S*G = S \cdot \Delta$.
(Neccesity.) From the fact that $S*G = S \cdot \Delta$ it follows that $Z(S) = C = Z(\Delta)$, and hence $Z(S) = Z(S)^G$. On the other hand, since Δ is Azumaya C-algebra, then Δ is H-separable over C. Let $\{d_i, v_i\}$ be an H-system for Δ over C, thus $d_i \in \Delta$ and $v_i \in C_{\Delta \otimes_C \Delta}(\Delta)$. But, because $S*G = S \cdot \Delta$, we can see v_i as element of $C_{S*G \otimes_S S*G}(S)$ under the natural homomorphism $\Delta \otimes_C \Delta \to S*G \otimes_S S*G$; giving us an H-system for $S*G$ over S. $\square$

In this case, the G-Galois condition in the commutator Δ is equivalent to an extra technical condition, which has been used in different ways to prove H-separability properties.

THEOREM 6. *Assume $S*G$ is H-separable over S and G acts trivially on the center of S; then Δ is G-Galois over C if and only if $\phi_g \phi_h = \phi_{hg}$ for all $g, h \in G$.*

Proof. (Sufficiency.) As G acts on Δ, let Φ_g be the corresponding ϕ_g. The commutator Δ being G-Galois over C, and hence over its center, is an Azumaya C-algebra. Thus, by [Kan65], $\Delta = \sum \oplus_{g \in G} \Phi_g$ and also $\Phi_g \Phi_{g^{-1}} = Z(\Delta)$. On the other hand, as a commutator of S in $S*G$, $\Delta = \sum \oplus_{g \in G} \phi_g g$. Let $t_g g \in \phi_g g$ and let $y \in \Delta$, then $(t_g g)^{g^{-1}} y = t_g g g^{-1} y g = t_g y g = y(t_g g)$. Hence $\phi_g g \subseteq \Phi_{g^{-1}}$. Equality follows because both are direct summands. Thus, $Z(\Delta) = \phi_g g \cdot \phi_{g^{-1}} g^{-1} = \phi_g {}^g \phi_{g^{-1}} = \phi_g \phi_{g^{-1}}$. It is clear that $\phi_g \phi_h \subseteq \phi_{hg}$, hence

$$\phi_{hg} \phi_{(hg)^{-1}} = Z(\Delta) = \phi_g(\phi_h \cdot \phi_{h^{-1}}) \phi_{g^{-1}} \subseteq \phi_g \phi_h(\phi_{g^{-1} h^{-1}} = \phi_g \phi_h(\phi_{(hg)^{-1}}$$

Therefore, $\phi_{hg} \subseteq \phi_g \phi_h$, but the reverse inclusion is trivial, thus $\phi_{hg} = \phi_g \phi_h$.
(Neccesity.) By theorem 5, Δ is an Azumaya C- algebra, therefore $\Delta \otimes_C \Delta \cong \mathrm{Hom}_C(\Delta, \Delta)$. Let j be the canonical homomorphism $\Delta * G \to \mathrm{Hom}_C(\Delta, \Delta)$, then

$$\Delta \otimes_C \Delta = \Delta \otimes \sum_{g \in G} \oplus \phi_g g = \sum_{g \in G} \oplus \Delta \otimes \phi_g g = \sum_{g \in G} \oplus j((\Delta \phi_g g)g^{-1}) = \sum_{g \in G} \oplus \Delta g^{-1} = \Delta * G.$$

Hence $\Delta * G \cong \mathrm{Hom}_C(\Delta, \Delta)$, and being Δ finitely generated as C-module, it is G-Galois . $\square$

REFERENCES

[Alf]   Ricardo Alfaro. Separabilities and *G*-Galois actions. to appear in Proceedings of the XXI Ohio State-Denison Conference.

[Alf92]   Ricardo Alfaro. Non-commutative separability and group actions. *Publicacions Matematiques*, 36:359–367, 1992.

[Hir69]   Kazuhiko Hirata. Separable extensions and centralizers of rings. *Nagoya Mathematical Journal*, 35:31–45, 1969.

[Ike81]   Shûichi Ikehata. Note on Azumaya algebras and H-separable extensions. *Mathematical Journal of Okayama University*, 23:17–18, 1981.

[Ike90]   Shûichi Ikehata. On H-separable polynomials of degree 2. *Mathematical Journal of Okayama University*, 32:53–59, 1990.

[Ike91]   Shûichi Ikehata. On H-separable polynomials of prime degree. *Mathematical Journal of Okayama University*, 33:21–26, 1991.

[Kan65]   Teruo Kanzaki. On Galois algebra over a commutative ring. *Osaka J. of Mathematics*, 2:309–317, 1965.

[NS75]   Taichi Nakamoto and Kozo Sugano. Note on H-separable extensions. *Hokkaido Mathematical Journal*, 4:295–299, 1975.

[OKI87]   Hiroaki Okamoto, Hiroaki Komatsu, and Shûichi Ikehata. On H-separable extensions in Azumaya algebras. *Mathematical Journal of Okayama University*, 29:103–107, 1987.

[OQ88]   James Osterburg and Declan Quinn. A Noether Skolem theorem for group-graded rings. *Journal of Algebra*, 113:483–490, 1988.

[Sug67]   Kozo Sugano. Note on semisimple extensions and separable extensions. *Osaka J. of Mathematics*, 4:265–270, 1967.

[Sug82]   Kozo Sugano. On H-separable extensions of two sided simple rings. *Hokkaido Mathematical Journal*, 11:246–252, 1982.

[Sug87]   Kozo Sugano. On H-separable extensions of primitive rings. *Hokkaido Mathematical Journal*, 16:207–211, 1987.

[Sug90]   Kozo Sugano. On H-separable extensions of primitive rings II. *Hokkaido Mathematical Journal*, 19:35–44, 1990.

Contributions of PI Theory to Azumaya Algebras

S. A. AMITSUR   Hebrew University, Jerusalem, Israel

To Prof. Dan Zelinsky for long friendship.

1. The famous Artin-Procesi theorem states that "An algebra A of rank n^2 is an Azumaya algebra over its center - if and only if - it satisfies all polynomial identities of the matrix ring $M_n(\mathbb{Z})$ over the integers $\mathbb{Z}$, and its simple homomorphic images do not satisfy identities of $M_{n-1}(\mathbb{Z})$". Later proofs show that its suffices to require that A satisfies the Cappelli identity $d_{n^2+1}[x, y] = 0$, and a certain central polynomial $g_n[x, y]$ of $M_n(\mathbb{Z})$ does not vanish on every simple image of A (e.g., [5] p. 66, [6] p. 100).

The "necessity" part, known as the "easy" part, since it follows from known properties of Azumaya algebras, but which require some rather complicated methods of reduction to the noetherization case and henselization. The "sufficiency" part, known as the "difficult" part has now some simple straightforward proofs (e.g., [6] p. 102).

Rowen has noticed in [5] (p. 65), that by using simple properties of localizations, and of PI-theory, one can obtain a fairly easy proof also to the "easy" part, and in fact to get new proofs of the properties of Azumaya algebras. He advocates using PI-theory to simplify some stages of the theory of Azumaya algebras, and, for example, he proves the existence of a splitting rings in the local case.

The theme of this work, is to push further this approach, and to obtain some of the basic properties of Azumaya algebras, avoiding the reduction to noether and henselization.

2. <u>up-central identities</u> Our interest centers around universal properties of matrices, which can be expressed by identities and central identities of matrices $M_n(\mathbb{Z})$.

A central polynomial $\varphi[x] = \varphi[x_1, \ldots, x_k]$ will be referred to as a *universally proper* (up-) central identity, if $\varphi[x]$ is a central polynomial in $M_n(\mathbb{Z})$, (hence in every $M_n(F)$, for any field F) and does not vanish in *every* simple algebra of dim n^2 over its center.

9

2.1. Examples of up-central identities have been given in [1, lemma 4]. Here we shall use the following identities derived from the cappelli identity $d_m[x, y] = 0$ defined as: (denoted $C_{2m}[x]$ in [5,6]):

(2.1) $d_m[x, y] = d[x_1, \ldots, x_m; y_1, \ldots, y_m] = \sum \text{sg} \sigma x_{\sigma(1)} y_1 x_{\sigma(2)} y_2 \cdots y_{m-1} x_{\sigma(m)} y_m$

where the sum ranges over all permutations.

2.2 The ring $M_n(\mathbb{Z})$ satisfies the cappelli identity $d_{n^2+1}[x, y] = 0$ and also a central identity $\delta_n[x, y] = \delta[x_1, \ldots, x_{n^2}; y_1, \ldots, y_{n^2+1}]$ (denoted by g_n in [5,6]) which is multilinear homogeneous and alternating in the x's only. Moreover $\delta_n[x, y]$ is up-central, since $\delta_n[M_n(H)] \neq 0$ in every $M_n(H)$, for arbitrary commutative H (e.g., [5] Theorem 1.4.17), and for any central simple algebra of dim n^2, $A \otimes H \cong M_n(H)$ for some H, hence $\delta_n[A] \neq 0$. Note also the alternativity of $\delta_n[x, y]$ will satisfy:

(2.2.0)   If $x_i' = \sum t_{ik} x_k$,

             then $\delta[x'; y] = \det(t_{ik}) \delta[x, y]$ for commutative $\{t_{ik}\}$.

This implies that a set $a = \{a_1, \ldots, a_{n^2}\}$ generates A (-is a base) if an only if there exist $b = \{b_i\}$ such that $\delta[a; b] \neq 0$ in A. Finally we recall that in any ring satisfying $d_{n^2+1}[x, y] = 0$ and $\delta_n[x; y]$ is central, we have for every u:

(2.2.1) $\delta[x; y]u = \sum_{i=1}^{n^2} \delta[x_1, \ldots, x_{i-1}, u, x_{i+1}, \ldots, x_{n^2}; y] x_i$

(e.g. (R2) in [3] p. 360), where this relation was used to obtain the following:

Let $x_1 = 1, x_2, \ldots, x_n, z; y_\lambda$ be a set of non commutative indeterminates, write

$\delta[x_i z^k; y] = \delta[1, z \ldots, z^{n-1}, x_2, x_2 z, \ldots, x_2 z^{n-1}, \ldots, x_n, x_n z, \ldots, x_n z^{n-1}; y]$

then by (R4) of [3] yields:

$\delta[x_j z^k; y] u x_i = \sum_{jk} \delta_{jk}[u x_i] x_j z^k$

(2.2.2)

$= \sum x_j \delta_{jk}[u x_i] z^k x = \sum_j x_j \tau_{jk}[u]$

where $\tau_{ji}[u] = \sum_k \delta_{jk}[u x_i] z^k$.

LEMMA 2.2.3. *Every central simple algebra A of dimension n^2 has a base $\{x_i z^j,\ i = 1, 2, \ldots, n,\ j = 0, 1, 2, \ldots, n - 1\}$ with $x_1 = 1$; such that also $\{1, z, \ldots, z^{n-1}, [x_i, z]z^j,\ 2 \leq i \leq n,\ 0 \leq j \leq n - 1\}$, where $[b, a] = ba - ab$,– is also a base of A over its center.*

PROOF: Consider first the well known case 1: $A = M_n(F)$, over an infinite field F:

Choose $z = \sum \zeta_i c_{ii}$ a diagonal matrix with $\zeta_i \neq \zeta_j$ for all $i \neq j$, and $x_j = x^{j-1}$, where $x = \sum\limits_{i=1}^{n} c_{ii+1}$, $(c_{nn+1} = c_{n1})$ and $\{c_{ik}\}$ be the matrix basis of $M_n(F)$. Then $xz = \sigma(z)x$, where $\sigma(z) = \zeta_2 c_{11} + \ldots + \zeta_n c_{n-1n-1} + \zeta_1 c_{nn}$. Thus, $\{z^i x^j\}$ form a base of $M_n(F)$; indeed, $x^\mu = \sum c_{\lambda\lambda+\mu}$, and there are polynomial $f_\lambda(t)$ of degree $< n$, such that $f_\lambda[\zeta_\mu] = \delta_{\lambda\mu}$, since the ζ_n's are different; thus $f_\lambda[z]x^\mu = c_{\lambda,\mu}$, and $f_\lambda[z] = c_{\lambda\lambda}$, and our assertion is easily concluded.

Moreover, $zx^j = x^j \sigma^j(z)$ and so $[z, x^j] = x^j(\sigma^{-j}(z) - z)$. The elements in the diagonal of z and of $z - \sigma^{-j}(z)$ are non zero for $j \geq 1$, hence we can find polynomials $h_j[t]$ of degree $< n$ such that $(\sigma^{-j}(z) - z)h_i[z] = z^i$. Thus $[z, x^j]h_i[z] = x^j z^i$, which implies that also $\{z^i, [z, x^j]z^i\}$ is also a base of $M_n(F)$, since $\{z^i x^j\}$ are a base.

This will prove our lemma for any central simple algebra A of dim n^2 over an infinite field. For, this implies that $\delta[Z^i X^j; Y] \neq 0$, $\delta[Z^i, [Z, X^j]Z^i; Y] \neq 0$ for generic matrices Z, X^j, Y, and hence by specialization it will follow also for the algebra A.

Case 2: If A is a simple algebra over a finite field F, then $A = M_n(F)$. In this case, F has a cyclic field extension K, with $(K : F) = n$, hence we can assume $M_n(F) \supset K$. The field K has a cyclic automorphism σ, which has an extension to an inner automorphism $\sigma(z) = xzx^{-1}$, where $K = F[z]$, and $x \in M_n(F)$. Finally, $M_n(F) = (K/F, \sigma, x^n)$ as a crossed-product, which implies that $\{x^j z^i\}$ and a base. Now, as before $\sigma^{-j}(z) - z \neq 0$ for $j > 1$, $[z, x^j] = x^j(\sigma^{-j}(z) - z)$ and $\sigma^{-j}(z) - z \neq 0$ in K. Choose $a_i = (\sigma^{-j}(z) - z)^{-1}z^i \in K = F[z]$, since $a_i = f_i(z)$ and we get $[z, x^j]f_i[z] = x^j z^i$ and thus $\{z^i, [z, x^j]z^i\}$ form base of $A = M_n(F)$.

REMARK 2.2.4: Note that in the proof (2.2.3) in both cases, we can choose z to be a separable element, i.e. all its characteristic roots be different.

COROLLARY 2.2.5. *The product of the up-central polynomials*

$$\gamma = \delta[z^i, x_j z^i, y]\delta[z^i, [z, x_j]z^i, y']$$

(y' a different set of indterminates) is up-central.

Indeed, substituting in any A central simple the elements z, x_j, found above, we obtain two non zero elements in the field-the center of A and hence also $\gamma \neq 0$.

2.3 Given the elements x_j, z as before and z satisfies its characteristic polynomial $z^n - c_1(z)z^{n-1} + \ldots + (-1)^n c_n = 0$, hence $\delta^{(i)}(z) := \delta[1, z, \ldots, z^{i-1}, z^n, z^{i+1}, \ldots, z^{n-1}, x_j \cdot z^i; y] = c_i(z)\delta[z^i, x_j z^i; y]$.

Now the discriminant of z, $d(z)$ is a homogeneous polynomial in the $c_i(z)$'s, we will get

$$(2.3.1) \qquad\qquad \delta^n[x_j z^i, y] \cdot d(z) = \eta[x, z, y]$$

and $\eta[x, z, y]$ is up-central, since there exist z separable and thus also $d[z] \neq 0$, as well as $\delta^n[x_j z^i, y]$.

Combining (2.2.4) and (2.3.1) we obtain similarly:

COROLLARY 2.3.2. *The product:*

$$\varphi[x, y, y', y'', z] = \delta[x_j z^i, y]\delta[z^i, [z, x_j]z^i, y']\eta[x, z, y'']$$

is a up-central polynomial (which we shall denote $\varphi[x, z]$.)

2.4. The last preparatory step is the following:

THEOREM 2.4.1. *If $\psi[x]$ is a up-central identity, which holds in A, and in all simple images of A then:*

(i) *There exists a finite set of substitutions, $x_i = a_{\lambda i} \in A$ such that $1 = \sum \psi[a_\lambda]$, $a_\lambda = \{a_{\lambda i}\}$,*

(ii) *Let $A_\lambda = A_{\psi[a_\lambda]}$ be the localization at $\psi[a_\lambda]$ of A, then $A \to \Pi A_\lambda$ is an injection, which maps also cent $(A) \to \Pi$ cent(A_λ).*

PROOF: Let $a \in \text{Ker}(A \to \Pi A_\lambda)$ then $\psi[a_\lambda]^m a = 0$ and we can choose the same m for all λ. Then clearly $1 \cdot a = (\sum \psi[a_\lambda])^k a = 0$ for some large k. The rest is evident.

3 APPLICATION. *Let A be an algebra which satisfies:*

(i) *The cappelli identity $d_{n^2+1}[x, y] = 0$*

(ii) *The polynomial $\delta_n[x; y]$ be central in A, and up-central in all its simple images.*

We shall write $\delta[x, y]$ and ommit the n. Condition (ii) is equivalent to

(ii)' *For every maximal ideal M in A, A/M is a central simple algebra of dimension n^2.*

Next theorem, which is a major tool in the theory of Azumaya algebras is proved here from basic properties of the polynomial identities of $M_n(\mathbb{Z})$:

THEOREM 3.1. *Let A satisfy (i), (ii), and $R = \text{cent}(A)$. If for some substitution $\varphi[x, y, z]$ in (2.3.3) is invertible in A, then*

a) *The element $z \in A$ is integral of degree n over the center R.*

b) *A is a free over $S = R[z]$ of dimension n.*

c) *$S = R[z]$ is a separable extension of R, and S is a maximal commutative subalgebra of A.*

d) *A is embedded in $M_n(S)$; moreover $A \underset{R}{\otimes} S \cong M_n(S)$ (i.e. S splits A). Hence A satisfies all polynomial identities of $M_n(\mathbb{Z})$.*

PROOF:

(a) From the definition of φ in corollary 2.3.2, it follows $\delta[z^i, x_j z^i, y] = \delta_1$ is also invertible in R as well as all the other factors of φ. Hence for $x_1 = 1$, $u = z^n$ we have by (2.2.1):

$$\delta_1 z^n = \sum_{j=1}^{n} x_j \varphi_i[z]$$

and $\varphi_j[z]$ are polynomials in $R[z]$ of degree $< n$. Now $\delta_1^{-1} \in R$, and so $z^n - \varphi_1[z] = \sum_{j=2}^{n} x_j \varphi_j[z]$. Taking the commutator $[z, -]$ on both sides yields $0 = \sum_{j=2}^{n} [z, x_j] \varphi_j[z]$ but by 2.3.3. we require that $\delta[z^i, [z, x_j] z^i, y'] \neq 0$, and so $\{[z, x_j] z^i\}$ are R-independent, hence all $\varphi_i[z] = 0$, and so $z^n - \varphi_1[z] = 0$, as required.

(b) To prove (b) we note that $\delta_1 \neq 0$ and invertible, hence if $\sum x_\nu \alpha_\nu[z] = \sum \alpha_{\nu\mu} x_\nu z^\mu = 0$ in R then for any fixed $\alpha_{\nu\mu}$, $\alpha_{\nu\mu} \delta[x_j z^i] = \delta[\ldots, \alpha_{\nu\mu} x_\nu z^\mu, \ldots] = 0$, since $\alpha_{\nu\mu} z^\nu x_\mu$ depends linearly on the rest of $x_j z^i$. But $\delta[x_j z^i] = \delta_1$ is invertible and so $\alpha_{\nu\mu} = 0$, and (b) is proved.

(c) $S = R[z]$ is separable, since $\eta[x, z, y''] \neq 0$ and invertible by our requirements (2.3.2). Therefore by (2.3.1) the discriminant of z is invertible and so $R[z]$ is a separable.

S is maximal commutative in A: Let $uz = zu$ for $u \in A$, then by (2.2.2) for $x_1 = 1$, we have

$$\delta_1[z^i x_j]u = \sum x_k \tau_k[u,z] = \tau_1[u,z] + \sum_{k=2}^{n} x_k \tau_k[u,z]$$

where $\tau_j[u,z]$ are polynomials in $R[z]$ of degree $\leq n - 1$. As before $\delta_1^{-1} \in R$, and $[z, x_k]z^i$ are a base, so commuting both sides with z yield $0 = \sum \delta_1^{-1} \tau_k[u,z][z, x_k]$ which implies that all $\tau_k[u,z] = 0$ and, therefore, $u = \delta_1^{-1} \tau_1[u,z] \in S$.

(d) It follows by (2.2.2), since $\delta_1 \neq 0$ and invertible, that the relation $ux_i = \sum x_j \delta_1^{-1} \tau_{ji}[u]$ yield a homomorphism $A \to M_n(S)$ by mapping $\pi : u \to (\delta_1^{-1}\tau_{ji}[u])$, and it is an injection, since $x_1 = 1$ and so if the image of $u = 0$, then $u = 0$.

This enables us to extend π to a homomorphism $\pi \otimes 1 : A \otimes S \to M_n(S)$, and we assert that $\pi \otimes 1$ is an isomorphism. Indeed $\pi[\delta[x_j z^i] = \delta[\pi(x_j)\pi(z)^i]$ is also invertible in $M_n(S)$ hence $\{\pi(x_j)\pi(z)^i\}$ are also S-generators of $M_n(S)$, i.e. $\sum \pi(x_j)\pi(z)^i S = M_n(S)$, which means that $\pi \otimes 1$ is-onto. Also $M_n(S)$ is free over the unit matrices $\{c_{ik}\}$, and if T is the transformation matrix which expresses $\{\pi(x_j)\pi(z)^i\}$ by the $\{c_{ik}\}$-then $\delta[\pi(x_j)\pi(z)^i] = (\det T)\delta[c_{ik}]$ by (2.2.0). The left hand side is the image of $\pi[\delta_1]$, and hence it is invertible so the same holds for T and, therefore, $\pi(x_j)\pi(z)^i$ are free generators over S. Consequently $\pi \otimes 1$ is an isomorphism.

Finally, the relation $A \underset{R}{\otimes} S \cong M_n(S)$, clearly yields that also A satisfies all identities of $M_n(\mathbb{Z})$.

Our main application is the basic property of the Azumaya algebras, which is now a consequences of the fact that algebras satisfying (i), (ii) are Azumaya algebras; but moreover it actually proves that they are Azumaya algebras.

THEOREM 3.2. *If A is an algebra satisfying* (i) *and* (ii), *then A has a splitting ring S, such that S is a faithfully flat extension of the center R of A. In fact S, is an étale covering of R, i.e. flat, separable and finitely presented.*

PROOF: If A satisfies (i) then the preceding arguments show that $\varphi[x,y,z]$ of theorem 3.1 is non trivial central identity in all the simple images of A.

It follows now by theorem (2.4.1) that there is a finite sum $\sum \varphi[x_\lambda, y_\lambda, z_\lambda] = 1$, and an embedding $A \to \Pi A_\lambda$. In each A_λ, theorem 3.1 is applicable and so A_λ has a splitting ring $Z_\lambda[z_\lambda] = S_\lambda$ which is free over Z_λ. Hence, $S = \Pi S_\lambda$ is the required splitting ring.

REMARK 3.3: By III.6.6. of [4], we in fact proved also that A is an Azumaya algebra.

3.4 Next we construct the reduced norm, the reduced trace and generally the characteristic polynomial for elements of the algebra A.

To this end we construct a generic element $a = \sum t_{ik} x_i z^k$ where $\{t_{ik}\}$ are n^2 commutative indeterminates, and $\{x_i z^k\}$ are the above set of generators for which $\delta = \delta[x_i z^k, y] \neq 0$, in the ring of polynomials $A_t = A[\ldots, t_{ik}, \ldots]$. Clearly A_t also satisfies (i) and (ii), and this follows also from (d) of (3.1) if δ is invertible, which implies that $A_t \otimes S_t \cong M_n(S_t)$ where $S_t = S[t_{ik}] = Z[z][t_{ik}]$.

Now let $a \in A_t \otimes S_t$, hence it annihilates its characteristic polynomial:

(3.4.1) $$f[\lambda; a] = \lambda^n - c_1(a)\lambda^{n-1} + \ldots + (-1)^n c_n(a)$$

$c_1(a) = \mathrm{tr}(a \otimes 1)$, $c_n(a) = \det(a \otimes 1)$; and all $c_i(a) \in S_k$.

LEMMA 3.4.2. If δ is invertible then all $c_i(a) \in R_t$, and independent of $1 \otimes z$.

PROOF: Let $c_i(a) = \sum_{\nu=0}^{n-1} \alpha_{i\nu}(1 \otimes z)^\nu$, $\alpha_{i\nu} \in R[t_{ik}] = R_t$, and put $c_0 = 1$. The fact that $f[a; a] = 0$ implies that

$$0 = \sum (-1)^u c_i(a) a^{n-i} = \sum_i \left(\sum_\nu \alpha_{i\nu}(a \otimes 1)^{n-i} \right)(1 \otimes z)^i.$$

Hence, $\sum_i \alpha_{i\nu} a^{n-i} = 0$ in A_t. But the generic element a will not satisfy a polynomial of degree $< n$ over R_t, since by specialization of the t's will get $a \to z$, and z is integral of degree n exactly. This shows that $\alpha_{i\nu} = 0$ for all $i > 0$ and all ν, since $\alpha_{i0} = 0$, and therefore $c_i(a) = \alpha_{i0} \in R_t$.

We turn now to the general case:

THEOREM 3.4.3. If A satisfies (i) and (ii), then A has a reduced trace function: $\mathrm{tr} : A \to R$, a reduced norm: Norm $A \to R$, and a characteristic polynomial: $f[\lambda; a] = \mathrm{Norm}(\lambda - a) \in R[\lambda]$, which have the same properties as for matrix ring; furthermore in any homomorphism $\psi : A \to M_n(K)$, K commutative, ψ maps traces, norms and characteristic polynomials of $u \in A$, onto the corresponding elements for $\psi(u)$ in $M_n(K)$.

PROOF: First we obtain this result for a universal algebra UA: Let $UA := \mathbb{Z}[X_i, Y_j, Z, U]$ the ring of generic matrices; and we localize this ring at the central element $\delta = \delta[X_i Z^k, Y_j]$. This ring $(UA)_\delta$ will clearly satisfy (i) and (ii), since $\delta = 0$ in every image of UA of pi-degree $< n$.

It follows now by (3.4.2), after specialization from a to U that $c_n(\delta U) = \det(\delta U) = \delta^n \det U$. On the other hand in the representation in (2.2.2) for UA, we have $\det(\delta U) = \det |\tau_{ij}[z]| \in \text{Cent}(UA)$. Thus it follows that:

$$(3.4.4) \qquad \delta^n \det U = \det(\delta U) = f[X, Y, Z, U] \in \text{Cent}(UA)$$

since δ is a non zero divisor also in UA (compare with [3]).

Next we pass to any algebra A satisfying (i) and (ii). Then, in any embedding, or even a homomorphism of A into a matrix ring $M_n(K)$, (3.4.4) will hold, as it is a relation between entries of the corresponding matrices.

Apply now theorem 2.4.1 to $\delta^n[x_i z^j, y]$ and we get $\sum c_\alpha \delta_\alpha^n = 1$, $c_\alpha \in R$, and the embedding $A \mapsto \Pi A_\alpha$. It follows now by (3.4.4)

$$(3.4.5) \qquad \det u = \sum c_\alpha \delta_\alpha^n \det u = g \in R$$

since $g = \sum c_\alpha f[x_\alpha, y_\alpha, z_\alpha, u_\alpha] \in \text{Cent}(A_\alpha) = R_\alpha$, as imbeddable in A_α and $M_n(S_\alpha) = A_\alpha \otimes S_\alpha$.

From (3.4.5), the characteristic polynomial is obtained by considering in $A[\lambda]$, the polynomial $\det |\lambda - u| \in R[\lambda]$ which is characteristic polynomial of u. This, and all the coefficients $c_j(u)$, will be mapped correspondingly, under any $\psi : A \to M_n(K)$, since $1 = \psi = (\sum c_\alpha \delta_\alpha^n) = \sum \psi(c_\alpha \delta_\alpha)^n$ in $M_n(K)$, and (3.4.4), (3.4.5) will be mapped, similarly, by ψ.

REMARK 3.4.6: These results are an additional step in fulfilling Rowen's request to contributions of PI-theory of Azumaya algebras. We recall that Rowen in [5] has shown how to get some more properties of Azumaya algebras straight from the identities.

REFERENCES

[1] S.A. Amitsur, Alternating Identities, Proceedings of Ohio conference on Ring Theory (Jain etc.), (1976) *Dekker*, pp. 1–14.

[2] S.A. Amitsur, Polynomial Identities and Azumaya Algebras, *Jour. Algebra*, 27: pp. 117–125 (1973).

[3] S.A. Amitsur, L.W. Small, Prime ideals in PI-rings, *Jour. Algebra*, 62: pp. 358–383 (1980).

[4] M.A. Knus, M. Ojanguren, Theorié de la Descrete et Algebre d'Azumaya', *Lecture notes in Mathematics*, 389 Springer, (1978)

[5] L.H. Rowen, Polynomial identities in Ring theory, *Pure and Applied Mathematics*, v. 84, Academic Press (1980).

[6] L.H. Rowen, Ring theory II, *Pure and Applied Mathematics*, v. 128, Academic Press (1988).

Cocycles and Right Crossed Products

M. BEATTIE, Department of Mathematics and Computer Science, Mount Allison University, Sackville, New Brunswick, Canada E0A 3C0

INTRODUCTION.

In [B2], a right crossed product is defined which generalizes the right smash product [D2], [B1], in the same way that the crossed product construction of Sweedler [S1], Doi and Takeuchi [DT], and Blattner, Cohen and Montgomery [BCM] generalizes the smash product $A \# H$ of a left H-module algebra A with a Hopf algebra H. The focus of [B2] was on duality results; if a classical crossed product is also a right crossed product then several technical assumptions in the duality theorems of Koppinen and Chen [K], [C], hold, and so duality statements hold for these crossed products.

As an aside in [B2], there is a short justification of the use of the word cocycle in the crossed product construction. In this note, we expand these remarks.

In §1 we define equivalence of cocycles and show that equivalent cocycles produce isomorphic crossed products, thus generalizing the commutative case where cocycles differing by a coboundary yield isomorphic crossed products. In §2 we show that although, of course, in general, the set of cocycles is not closed under multiplication, some cocycles do fit neatly into an algebraic structure. We define a $Z/2Z$-graded ring whose homogeneous components are cocycles in $1 \otimes H \otimes H \subset A \otimes H \otimes H$ where $A = H$ or $A = H^{cop}$. If (H, R) is a quasitriangular Hopf algebra, $1 \otimes R$ is a degree 1 homogeneous element of this ring.

§0. PRELIMINARIES.

We work over a commutative ring k with 1; all maps are k-linear. Throughout, H will denote a k-bialgebra; we use Sweedler's summation notation [S2]. We will write $\Delta^2(h)$ for $\Delta \otimes I \cdot \Delta(h) = I \otimes \Delta \cdot \Delta(h)$.

A k-algebra A is said to have a weak right H-coaction if there is a linear map $\alpha : A \to A \otimes H$ such that α is an algebra homomorphism and $(I \otimes \epsilon) \cdot \alpha = I$. We write

$$\alpha(a) = \sum a^0 \otimes a^1, \ \alpha^2(a) = \alpha \otimes I \cdot \alpha(a) = \sum a^0 \otimes a^1 \otimes a^2$$

for $a \in A$. Note that $\alpha^2(a)$ may not equal $(I \otimes \Delta) \cdot \alpha(a)$. Throughout, A will denote an algebra with weak right H-coaction and structure map α.

If H, A are commutative and A is a right H-comodule (i.e. here $\alpha^2(a) = I \otimes \Delta \cdot \alpha(a)$), then a cochain complex $\mathcal{C}(A, H, \mathcal{U})$ can be defined where $\mathcal{U}$ is the units functor. Let $H^0 = k$, $H^n = H \otimes H \otimes \ldots \otimes H$, n copies of H, for $n > 0$. For $n \geq 1$, define H-comodule algebra maps $\Delta_i, 0 \leq i \leq n$, from $A \otimes H^{n-1}$ to $A \otimes H^n$ by

$$\Delta_0 = \alpha \otimes I \otimes \ldots \otimes I,$$

$$\Delta_i = I \otimes \ldots \otimes \Delta_H \otimes \ldots \otimes I, \Delta_H \text{ acts on the } i\text{th copy of } H, 1 \leq i \leq n - 1,$$

$$\Delta_n = I \otimes \ldots \otimes I \otimes \mu_H, \mu_H \text{ the unit map from } k \text{ to } H.$$

Let $C^n(A, H, \mathcal{U}) = \mathcal{U}(A \otimes H^n), n \geq 0$, with $\delta_n : C^{n-1} \to C^n$ defined by $\delta_n = \Pi_{i=0}^n \mathcal{U}(\Delta_i)^{(-1)^i}$. If $H = kG$, G an abelian group and $A = k$, this complex yields the usual Harrison cohomology [CR, p.664]. If H is a finitely generated projective Hopf algebra, this complex yields the cohomology $H^*(\text{Hom}(H, k), A)$ of [S1].

Let $s = \sum a_i \otimes h_i \otimes g_i \in A \otimes H \otimes H$. We say that s satisfies the twisted comodule condition (TC) if, for all $c \in A$, $(\alpha \otimes I \cdot \alpha(c))s = s(I \otimes \Delta \cdot \alpha(c))$, i.e.,

$$\sum c^{(0)}a_i \otimes c^{(1)}h_i \otimes c^{(2)}g_i = \sum a_i c^{(0)} \otimes h_i c^{(1)}{}_{(1)} \otimes g_i c^{(1)}{}_{(2)}. \tag{TC}$$

We say s satisfies the cocycle condition (C) if $\Delta_0(s)\Delta_2(s) = \Delta_3(s)\Delta_1(s)$, i.e.

$$\sum a_i^{(0)}a_j \otimes a_i^{(1)}h_j \otimes h_i g_{j_{(1)}} \otimes g_i g_{j_{(2)}} = \sum a_k a_t \otimes h_k h_{t_{(1)}} \otimes g_k h_{t_{(2)}} \otimes g_t. \tag{C}$$

If s satisfies (TC) and (C), we say s is a cocycle (for A). If H is commutative and A is a right H-comodule, then (TC) is trivially satisfied and (C) is equivalent to $\delta_3(s) = 1$.

If L is a right H-module ring, we define the right crossed product $A\#_s L$ to be the k-module $A \otimes L$ with multiplication given by

$$(a\#_s l)(b\#_s m) = \sum ab^0 a_i \#_s (l \leftharpoonup b^1 h_i)(m \leftharpoonup g_i).$$

If s is a cocycle, then this multiplication is associative and if s is a normal cocycle, i.e., if

$$(I \otimes \epsilon \otimes I)(s) = (I \otimes I \otimes \epsilon)(s) = 1 \otimes 1 \in A \otimes H, \qquad \text{i.e.,}$$

$$\sum a_i \epsilon(h_i) \otimes g_i = \sum a_i \epsilon(g_i) \otimes h_i = 1 \otimes 1, \tag{N}$$

then $1\#_s 1$ is the multiplicative identity in $A\#_s L$. If A is an H-comodule and $s = 1 \otimes 1 \otimes 1$, then $A\#_s L = A\#L$, the right smash product [B1].

Recall [R, §2] that (H, R), H a Hopf algebra, $R = \sum R^{(1)} \otimes R^{(2)} \in H \otimes H$, is called a quasitriangular Hopf algebra if the following conditions hold $(r = R)$:

(QT.1) $\sum \Delta(R^{(1)}) \otimes R^{(2)} = \sum R^{(1)} \otimes r^{(1)} \otimes R^{(2)}r^{(2)}$,

(QT.2) $\sum \epsilon(R^{(1)})R^{(2)} = 1$,

(QT.3) $\sum R^{(1)} \otimes \Delta(R^{(2)}) = \sum R^{(1)}r^{(1)} \otimes r^{(2)} \otimes R^{(2)}$

(QT.4) $\sum R^{(1)}\epsilon(R^{(2)}) = 1$,

(QT.5) $\triangle^{cop}(h)R = R\triangle(h)$ for all $h \in H$.

Elements R with these properties give examples of the cocycles we study in §3.

§1. EQUIVALENT COCYCLES AND RIGHT CROSSED PRODUCTS.

In this section, we introduce an equivalence relation on cocycles and show that equivalent cocycles yield isomorphic crossed products. If H, A are commutative, cocycles are equivalent if and only if they differ by a coboundary. Our results are analogous to those in [D1] and [BCM] for the classical crossed product.

Throughout this section, A will denote a k-algebra with weak H-coaction $\alpha : A \to A \otimes H$. The same algebra A will be written A' or A'' if it is equipped with a different weak coaction map α' or α''. We will denote the algebra with trivial coaction by $|A|$; here the coaction map is the inclusion map i, $i(a) = a \otimes 1$.

Let $s \in A \otimes H \otimes H$, $t \in A' \otimes H \otimes H$. We say s, t are equivalent if there exists $w \in \mathcal{U}(A \otimes H)$ such that

(i) $\alpha(a) = w\alpha'(a)w^{-1}$,   and

(ii) $s = (\alpha \otimes I)(w)(w \otimes 1)t\triangle_1(w^{-1})$

$= (w \otimes 1)(\alpha' \otimes I)(w)t\triangle_1(w^{-1})$ by (i).

We write $s \sim t$ (via w).

LEMMA 1.1. The relation $\sim$ above is an equivalence relation.

Proof. Clearly $s \sim s$ via $1 \otimes 1$. Also if $s \sim t$ via w as above, $t \sim s$ via w^{-1}. Suppose $s \sim t$ via w and $t \sim p$ via u with $p \in A'' \otimes H \otimes H$. Then

$$\begin{aligned}
\alpha(a) &= w\alpha'(a)w^{-1} = (wu)\alpha''(a)(wu)^{-1}, \text{ and} \\
s &= (w \otimes 1)(\alpha' \otimes I)(w)t\triangle_1(w^{-1}) \\
&= (w \otimes 1)(\alpha' \otimes I)(wu)(u \otimes 1)p\triangle_1(wu)^{-1} \\
&= (\alpha \otimes I)(wu)(wu \otimes 1)p\triangle_1(wu)^{-1}.
\end{aligned}$$

Thus $s \sim p$ via wu.   ∎

Now we prove the main theorem of this section.

THEOREM 1.2. Let $s \in A \otimes H \otimes H$, $t \in A' \otimes H \otimes H$ and $s \sim t$ via w. Then

(i) For all right H-module rings L, $A\#_s L \cong A'\#_t L$.

(ii) The element s satisfies (TC) for A if and only if t satisfies (TC) for A'.

(iii) If s satisfies (TC) for A, then s is a cocycle if and only if t is also.

(iv) If $(I \otimes \epsilon \otimes I)(t)$ $((I \otimes I \otimes \epsilon)(t))$ is $1 \otimes 1$, then $(I \otimes \epsilon \otimes I)(s)$ $((I \otimes I \otimes \epsilon)(s))$ is also if and only if $(I \otimes \epsilon)(w) = 1$. Also if $(I \otimes \epsilon)(w) = 1$, then $(I \otimes \epsilon \otimes I)(t)$ $((I \otimes I \otimes \epsilon)(t))$ is $1 \otimes 1$ if and only if $(I \otimes \epsilon \otimes I)(s)$ $((I \otimes I \otimes \epsilon)(s))$ is also, so that if $(I \otimes \epsilon)(w) = 1$, s satisfies (N) if and only if t does.

Proof. Suppose $s = \sum a_i \otimes h_i \otimes g_i \in A \otimes H \otimes H$, $t = \sum b_j \otimes l_j \otimes m_j \in A' \otimes H \otimes H$ and $w = \sum c_j \otimes y_i$, $w^{-1} = \sum d_k \otimes z_k \in A \otimes H$. Then $t = (w^{-1} \otimes 1)(\alpha \otimes I)(w^{-1})s\Delta_1(w)$.

(i) Let $\phi : A\#_s L \to A\#_t L$ be defined by $\phi(a\#_s l) = \sum ac_i \#_t l \leftharpoonup y_i$. Clearly ϕ is a bijection; we must show it preserves multiplication.

For $a, b \in A$, $l, m \in L$, we have

$\phi[(a\#_s l)(b\#_s m)]$

$= \phi[\sum ab^0 a_i \#_s(l \leftharpoonup b^1 h_i)(m \leftharpoonup g_i)]$

$= \sum ab^0 a_i c_k \#_t(l \leftharpoonup b^1 h_i)(m \leftharpoonup g_i) \leftharpoonup y_k$

$= \sum ab^0 c_p^0 c_l b_j \#_t(l \leftharpoonup b^1 c_p^1 y_l l_j)(m \leftharpoonup y_p m_j)$

$= \sum ac_l(b'c_p')^0 b_j \#_t(l \leftharpoonup y_l(b'c_p')^1 l_j)(m \leftharpoonup y_p m_j)$ where $\alpha'(b) = b'^0 \otimes b'^1$

$= \sum(ac_l \#_t l \leftharpoonup y_l)(bc_p \#_t m \leftharpoonup y_p)$

$= \phi(a\#_s l)\phi(b\#_s m)$ as required.

(ii) Suppose s satisfies (TC), i.e. for all $c \in A$,

$$\alpha^2(c)(s) = (s)(I \otimes \Delta) \cdot \alpha(c).$$

Thus $\alpha^2(c)(s)\Delta_1(w) = (s)\Delta_1(w)(I \otimes \Delta) \cdot \alpha'(c)$. If we multiply both sides of this equality on the left by $(w^{-1} \otimes 1)(\alpha \otimes I)(w^{-1})$, the right hand side becomes $t(I \otimes \Delta)\alpha'(c)$ and the left hand side is

$$(w^{-1} \otimes 1)(\alpha \otimes I)(w^{-1})\alpha^2(c)(s)\Delta_1(w)$$

$$= (w^{-1} \otimes 1)(\alpha \otimes I)[w^{-1}\alpha(c)]s\Delta_1(w)$$

$$= (w^{-1} \otimes 1)(\alpha \otimes I)[\alpha'(c)w^{-1}]s\Delta_1(w)$$

$$= (w^{-1} \otimes 1)(w \otimes 1)(\alpha' \otimes I)[\alpha'(c)w^{-1}](w^{-1} \otimes 1)s\Delta_1(w)$$

$$= \alpha'^2(c)t \text{ as required.}$$

(iii) $\Delta_0(t)\Delta_2(t)(I \otimes \Delta^2)(w^{-1})$

$$= \sum(\alpha' \otimes I \otimes I)(t)(w^{-1} \otimes 1 \otimes 1)(\alpha \otimes \Delta)(w^{-1})\Delta_2(s)$$

$$= \sum(w^{-1} \otimes 1 \otimes 1)(\alpha \otimes I \otimes I)(t)(\alpha \otimes \Delta)(w^{-1})\Delta_2(s)$$

$$= \sum(w^{-1} \otimes 1 \otimes 1)(\alpha \otimes I \otimes I)(w^{-1} \otimes 1)(\alpha^2 \otimes I)(w^{-1})\Delta_0(s)\Delta_2(s)$$

Also $\Delta_3(t)\Delta_1(t)(I \otimes \Delta^2)(w^{-1})$
$= \sum(w^{-1} \otimes 1 \otimes 1)((\alpha \otimes I)w^{-1} \otimes 1)\Delta_3(s)((I \otimes \Delta) \cdot \alpha \otimes I)(w^{-1})(w^{-1})\Delta_1(s)$
$= \sum(w^{-1} \otimes 1 \otimes 1)((\alpha \otimes I)w^{-1} \otimes 1)(\alpha^2 \otimes I)(w^{-1})\Delta_3(s)\Delta_1(s)$ by (TC).
Thus it is clear that $\Delta_0(t)\Delta_2(t) = \Delta_3(t)\Delta_1(t)$ if and only if $\Delta_0(s)\Delta_2(s) = \Delta_3(s)\Delta_1(s)$.

(iv) Since $(I \otimes \epsilon \otimes I)(s) = (I \otimes \epsilon)(w)w(I \otimes \epsilon \otimes I)(t)w^{-1}$, then the statements for the map $I \otimes \epsilon \otimes I$ are clear. Also $(I \otimes I \otimes \epsilon)(s) = \alpha \cdot (I \otimes \epsilon)(w)w(I \otimes I \otimes \epsilon)(t)w^{-1}$, and we need only note that $(I \otimes \epsilon)(w) = 1$ if and only if $\alpha \cdot I \otimes \epsilon(w) = 1 \otimes 1$. Now the statements for $I \otimes I \otimes \epsilon$ are also straightforward. ∎

COROLLARY 1.3. If $s \in A \otimes H \otimes H$ and $w \in \mathcal{U}(A \otimes H)$ commutes with $\alpha(A)$, then $s \sim t$ via w where

$$t = \Delta_0(w)\Delta_2(w)s\Delta_1(w) \in A \otimes H \otimes H.$$

In particular, [B2, 1.11 (i)], if $s = 1 \otimes 1 \otimes 1$, $t = \Delta_0(w)\Delta_2(w)\Delta_1(w)$,

$$A\#_t L \cong A\#L.$$

∎

Recall [BCM, 2.2] that a weak H-coaction on an algebra A is called inner if there is a $w \in \mathcal{U}(A \otimes H)$ such that $\alpha(a) = w(a \otimes 1)w^{-1}$ for all $a \in A$. We say w implements the inner weak coaction.

COROLLARY 1.4. If α is an inner weak coaction implemented by w, then for any $s \in A \otimes H \otimes H$, $s \sim t$ via w where

$$
\begin{aligned}
t &= (i \otimes I)(w^{-1})(w^{-1} \otimes 1)s\Delta_1(w) \\
&= (w^{-1} \otimes 1)(\alpha \otimes I)(w^{-1})s\Delta_1(w) \in |A| \otimes H \otimes H.
\end{aligned}
$$

Thus $A\#_s L \cong |A|\#_t L$ where the right hand side is analogous to the twisted product of [BCM, 5.3], [D1, 2.3]. In particular, if $s = \Delta_0(w)\Delta_2(w)\Delta_1(w^{-1})$, then $A\#_s L \cong A \otimes L$ since then $t = 1$, [B, 1.11 (ii)]. ∎

REMARK 1.5. Recall that an inner weak coaction is called strongly inner if it is implemented by an element $w \in \mathcal{U}(A \otimes H)$ such that $\Delta_1(w) = \Delta_0(w)\Delta_2(w)$. In this case, then, we obtain the well-known fact that $A\#L \cong A \otimes L$ for all L. ∎

§2. MULTIPLICATION OF COCYCLES FROM $1 \otimes H \otimes H$.

Throughout this section H^{cop} will denote the bialgebra which equals H as an algebra but with opposite comultiplication Δ^{cop}. The twist map will be denoted tw and we write s^{tw} for $tw(s)$. Then $\Delta^{cop} = tw \cdot \Delta$.

In general, for A and H not commutative, the product of cocycles in $A \otimes H \otimes H$ need not be a cocycle. However, the product of cocycles of the form $1 \otimes s \in 1 \otimes H \otimes H \subset H \otimes H \otimes H$ is again such a cocycle. Condition (TC) is just the condition that s commutes with $\Delta(H) \subset H \otimes H$. Then if $1 \otimes s, 1 \otimes t$ in $1 \otimes H \otimes H$, satisfy (C), since (TC) guarantees that $\Delta_0(1 \otimes t)$, $\Delta_2(1 \otimes s)$ commute and $\Delta_3(1 \otimes t)$, $\Delta_1(1 \otimes s)$ commute,

$$
\begin{aligned}
\Delta_0(1 \otimes st)\Delta_2(1 \otimes st) &= \Delta_0(1 \otimes s)\Delta_2(1 \otimes s)\Delta_0(1 \otimes t)\Delta_2(1 \otimes t) \\
&= \Delta_3(1 \otimes s)\Delta_1(1 \otimes s)\Delta_3(1 \otimes t)\Delta_1(1 \otimes t) \\
&= \Delta_3(1 \otimes st)\Delta_1(1 \otimes st)
\end{aligned}
$$

so that (C) holds for $1 \otimes st$.

We note that if $1 \otimes s \in H \otimes H \otimes H$ is a cocycle, then $1 \otimes s \in A \otimes H \otimes H$ is a cocycle for any right H-comodule algebra A.

Now consider $s \in H \otimes H$ such that $1 \otimes s \in H^{cop} \otimes H \otimes H$ is a cocycle, and therefore a cocycle in $A \otimes H \otimes H$ for every right H^{cop}-comodule algebra A. The product of these cocycles need not be a cocycle. Our goal in this section is to build an algebraic structure which will incorporate both types of cocycles above.

DEFINITION 2.1 (i) $C(H, H) = \{s | s \in H \otimes H, 1 \otimes s \text{ is a cocycle in } H \otimes H \otimes H\}$.

(ii) $C(H, H^{cop}) = \{s | s \in H^{cop} \otimes H^{cop}, 1 \otimes s \text{ is a cocycle in } H \otimes H^{cop} \otimes H^{cop}\}$.

(iii) $C(H^{cop}, H) = \{s | s \in H \otimes H, 1 \otimes s \text{ is a cocycle in } H^{cop} \otimes H \otimes H\}$.

(iv) $C(H^{cop}, H^{cop}) = \{s | s \in H^{cop} \otimes H^{cop}, 1 \otimes s \text{ is a cocycle in } H^{cop} \otimes H^{cop} \otimes H^{cop}\}$. ∎

For example, $s \in H^{cop} \otimes H^{cop}$ lies in $C(H, H^{cop})$ if and only if for all $h \in H$,

$$\Delta(h)s = s\Delta^{cop}(h), \text{ and } (1 \otimes s)(1 \otimes \Delta^{cop})(s) = (s \otimes 1)(\Delta^{cop} \otimes 1)(s).$$

We saw above that $C(H, H)$ is closed under multiplication; so is $C(H^{cop}, H^{cop})$.

Elements in $H \otimes H$ satisfying QT conditions (see §0) lie in $C(H^{cop}, H)$. For (QT5) for $R \in H \otimes H$ holds if and only if we have (TC) for $1 \otimes R \in H^{cop} \otimes H \otimes H$. Also if $R \in H \otimes H$ satisfies (QT1) and (QT3), then (C) holds for $1 \otimes R$ if and only if

$$R^{\hat{1}} R^{\hat{2}} R^{\hat{3}} = R^{\hat{3}} R^{\hat{2}} R^{\hat{1}},$$

i.e. if and only R is a solution to the quantum Yang-Baxter equation. (See [R, p.290].) If (QT5) holds, then it is. As well, $1 \otimes R$ satisfies (N) if and only if (QT2) and (QT4) hold.

LEMMA 2.2 Suppose $s \in C(H, H^{cop}), t \in C(H^{cop}, H), u \in C(H, H), w \in C(H^{cop}, H^{cop})$. Then $st \in C(H, H)$, $ts \in C(H^{cop}, H^{cop})$, us and sw are in $C(H, H^{cop})$, and wt and tu are in $C(H^{cop}, H)$.

Proof. We show $st \in C(H, H)$; the rest of the proof is similar. For all $h \in H$,

$$st(\Delta h) = s(\Delta^{cop}h)t = \Delta(h)st,$$

so (TC) holds for st. Also $\Delta_0(1 \otimes st)\Delta_2(1 \otimes st) = \Delta_3(1 \otimes st)\Delta_1(1 \otimes st)$ since

$$
\begin{aligned}
(1 \otimes s)(1 \otimes t)(I \otimes \Delta)(s)(I \otimes \Delta)(t) &= (1 \otimes s)(I \otimes \Delta^{cop})(s)(1 \otimes t)(I \otimes \Delta)(t) \\
&= (s \otimes 1)(\Delta^{cop} \otimes I)(s)(t \otimes 1)(\Delta \otimes I)(t) \\
&= (s \otimes 1)(t \otimes 1)(\Delta \otimes I)(s)(\Delta \otimes I)(t). \blacksquare
\end{aligned}
$$

LEMMA 2.3 The twist map tw induces bijections between $C(H, H^{cop})$ and $C(H^{cop}, H)$ and between $C(H, H)$ and $C(H^{cop}, H^{cop})$.

Proof. We show that the twist map is a bijection from $C(H, H^{cop})$ to $C(H^{cop}, H)$; the second argument is similar. Note that for all $h \in H, s \in H \otimes H$,

$$\Delta(h)s = s\Delta^{cop}(h) \text{ if and only if } \Delta^{cop}(h)s^{tw} = s^{tw}\Delta(h),$$

so $1 \otimes s \in H \otimes H^{cop} \otimes H^{cop}$ satisfies (TC), if and only if $1 \otimes s^{tw} \in H^{cop} \otimes H \otimes H$ does too. Also (C) for $1 \otimes s \in H \otimes H^{cop} \otimes H^{cop}$ holds if and only if

$$(1 \otimes s)(I \otimes \Delta^{cop})(s) = (s \otimes 1)(\Delta^{cop} \otimes I)(s).$$

This holds if and only if $(tw \otimes I)(I \otimes tw)(tw \otimes I)$ applied to both sides of the above is an equality, i.e. if

$$(s^{tw} \otimes 1)(\Delta \otimes I)(1 \otimes s^{tw}) = (1 \otimes s^{tw})(I \otimes \Delta)(s^{tw}).$$

But this holds if and only if (C) holds for $1 \otimes s^{tw} \in H^{cop} \otimes H \otimes H$. $\blacksquare$

Now we define a $Z/2Z$-graded ring whose homogeneous components are the cocycles in $C(H, H)$ and $C(H^{cop}, H)$.

PROPOSITION 2.4. Let C be the k-module $C(H, H) \oplus C(H^{cop}, H), Z/2Z$-graded by $C_0 = C(H, H), C_1 = C(H^{cop}, H)$. Then C with multiplication m is a $Z/2Z$-graded ring where m is given by $m(s, t) = st$, the usual product in $H \otimes H$, if $t \in C_0$ and $m(s, t) = (s^{tw})t$ if $t \in C_1$.

Proof. It is clear that multiplication behaves well with respect to grading and it remains to show that multiplication is well-defined and associative. This can be seen by straightforward computation of the possible cases. For example if $s, u \in C_1$ and $t \in C_0$, then note that $st \in C(H^{cop}, H), (st)^{tw} \in C(H, H^{cop})$ so that $(st)^{tw}u \in C(H, H), t^{tw} \in C(H^{cop}, H^{cop})$ so that $t^{tw}u \in C(H^{cop}, H)$ and

$s^{tw}(t^{tw}u) \in C(H,H)$. Thus $m(m(s,t),u)$ and $m(s,m(t,u))$ are defined and it is easily checked that both equal $(st)^{tw}u$. The remaining cases are equally straightforward. ∎

Acknowledgement

This work was supported in part by NSERC grant OGP0009137. Thanks to W. Chin for pointing out the connection between (C) and the quantum Yang-Baxter equation.

References

[B1] M. Beattie, On the Blattner-Montgomery duality theorem for Hopf algebras, in Contemporary Mathematics, 124, 23-28, A.M.S., Providence, R.I. 1992.

[B2] M. Beattie, A right crossed product and duality, Comm. Alg., to appear.

[BCM] R. J. Blattner, M. Cohen and S. Montgomery, Crossed products and inner actions of Hopf algebras, Trans. Amer. Math. Soc. 298 (1986), 671-711.

[CR] S. U. Chase and A. Rosenberg, A theorem of Harrison, Kummer theory and Galois algebras, Nagoya Math. J. 27 (1966), 663-685.

[C] Chen Caoyu, A duality theorem for crossed products of Hopf algebras, Comm. Alg., to appear.

[D1] Y. Doi, Equivalent crossed products for a Hopf algebra, Comm. Algebra 17 (1989), 3053-3085.

[D2] Y. Doi, Unifying Hopf modules, J. Algebra 153 (1992), 373-385.

[DT] Y. Doi and M. Takeuchi, Cleft comodule algebras for a bialgebra, Comm. Algebra 14 (1986), 801-818.

[K] M. Koppinen, A duality theorem for crossed products of Hopf algebras, J. Algebra 146 (1992), 153-174.

[R] D. E. Radford, Minimal quasitriangular Hopf algebras, J. Algebra 157 (1993), 285-315.

[S1] M. E. Sweedler, Cohomology of algebras over Hopf algebras, Trans. Amer. Math. Soc. 133 (1968), 205-239.

[S2] M. E. Sweedler, "Hopf algebras", Benjamin, New York, 1969.

Engel-type Theorems for Lie Color Algebras

JEFFREY BERGEN   Department of Mathematics, DePaul University, Chicago, Illinois 60614

PIOTR GRZESZCZUK   University of Warsaw, Białystok Division, Akademicka 2,15-267, Białystok, Poland

At present, there is a great deal of work concerned with the actions of Hopf algebras on associative rings [7]. Much of this work has centered on the cases where the Hopf algebra is a group algebra KG [6], the dual of a group algebra $(KG)^*$ [2], or the enveloping algebra $U(L)$ or restricted enveloping algebra $u(L)$ of a Lie algebra [1], [5]. KG, $U(L)$, and $u(L)$ are cocommutative Hopf algebras, whereas $(KG)^*$ is a commutative Hopf algebra. However, recent work on quantum groups has fueled the interest in Hopf algebras which are neither commutative nor cocommutative. A difficulty in studying the actions of Hopf algebras which are neither commutative nor cocommutative has been a lack of concrete examples.

It was shown in [3] that given a Lie color algebra L, its enveloping algebra or restricted enveloping algebra is contained, in a natural way, in a noncommutative, noncocommutative Hopf algebra. In order to study the actions of this concrete class of noncommutative, noncocommutative Hopf algebras, one needs to first examine the modules over Lie color algebras. The purpose of this paper is to present some facts, similar to Engel's theorem for Lie algebras, about Lie color algebras. These results may be known, but we do not know if they have previously appeared in print. However, since these results may be helpful in the study of the actions of this class of noncommutative, noncocommutative Hopf algebras, we present them here. The arguments we use are motivated by those of Jacobson in the proof of Theorem 11 in [4].

In all that follows, L will be a vector space over a field K of characteristic different from 2. If G is an abelian group, we call a map $\epsilon : G \times G \longrightarrow K^*$ a bicharacter if $\epsilon(g, hk) = \epsilon(g, h)\epsilon(g, k)$, $\epsilon(gh, k) = \epsilon(g, k)\epsilon(h, k)$, and $\epsilon(g, h) = \epsilon(h, g)^{-1}$ for all $g, h, k \in G$. L is said to be a G-graded algebra if $L = \bigoplus_{g \in G} L_g$ and L has a multiplication $[,]$ such that $[L_g, L_h] \subseteq L_{gh}$ for all $g, h \in G$. Next we say that L is a color Lie algebra if L is a G-graded algebra and there exists a bicharacter ϵ such that $[x, y] = -\epsilon(g, h)[y, x]$ and $[[x, y], z] = [x, [y, z]] - \epsilon(g, h)[y, [x, z]]$ for all $x \in L_g$, $y \in L_h$, and $z \in L$. The elements of $\bigcup_{g \in G} L_g$ are known as the homogeneous elements of L. We will say that L is nilpotent if there exists a positive integer M such that $[x_M, [x_{M-1}, [\cdots, [x_2, x_1] \cdots]]] = 0$ for all $x_1, \ldots, x_M \in L$.

31

A K-vector space V is an L-module if there is a vector space homomorphism $\psi : L \longrightarrow End_K(V)$ such that $\psi([x,y]) = \psi(x)\psi(y) - \epsilon(g,h)\psi(y)\psi(x)$ for all $x \in L_g$ and $y \in L_h$. Our main result will follow from

PROPOSITION 1. *Let A be a finite-dimensional subspace of $End_K(V)$ spanned by a set S such that for every $a,b \in S$ there exists $\alpha = \alpha(a,b) \in K^*$ such that $ab - \alpha ba \in S$. If every element of S is a nilpotent transformation of V, then A acts nilpotently on V. That is, there exists an integer N such that the composition of any N transformations from A is 0.*

Proof. For any subset B of A, we will let $sp(B)$ denote the subspace of A spanned by B. Consider all subsets T of S with the properties that T acts nilpotently on V and for every $a,b \in T$, there exists $\alpha = \alpha(a,b) \in K^*$ such that $ab - \alpha ba \in sp(T)$. Without loss, we may assume that S contains 0, therefore such subsets certainly exist. From among all these subsets, choose one, U, such that the dimension of $sp(U)$ is maximal. It suffices to show that $sp(U) = A$.

If $a, a_1, a_2, \ldots, a_n \in S$ we can inductively define $F_0(a) = a$, $F_1(a, a_1) = aa_1 - \alpha(a, a_1)a_1a$, and if $t = F_{n-1}(a, a_1, a_2, \ldots, a_{n-1})$, then $F_n(a, a_1, a_2, \ldots, a_n) = F_1(t, a_n)$. If m is an integer such that $U^m = 0$; let $a \in S$, $u_1, \ldots, u_{2m-1} \in U$, and consider $F_{2m-1}(a, u_1, \ldots, u_{2m-1})$. Expanding $F_{2m-1}(a, u_1, \ldots, u_{2m-1})$ in $End_K(V)$ results in a linear combination of products of elements of S, where each product contains a string of at least m consecutive elements from U. Therefore $F_{2m-1}(a, u_1, \ldots, u_{2m-1}) = 0$ and so, $F_{2m-1}(S, U, \ldots, U) = 0$.

If $sp(U) \neq A$, let l be the smallest integer such that $F_l(S, U, \ldots, U) \subseteq sp(U)$. Since $l \geq 1$, there exist $a \in S$ and $a_1, \ldots, a_{l-1} \in U$ such that $F_{l-1}(a, a_1, \ldots, a_{l-1}) \notin sp(U)$. Let $c = F_{l-1}(a, a_1, \ldots, a_{l-1})$; then $F_1(c, U) \subseteq F_l(S, U, \ldots, U) \subseteq sp(U)$. Now consider the set $U_1 = U \cup \{c\}$. By the construction of c, for any $u \in U$ there exists $\beta \in K^*$ such that $cu - \beta uc \in sp(U)$. Since β is invertible, it follows that $uc - \beta^{-1}cu \in sp(U)$. Thus for any $a, b \in U_1$, there exists $\alpha \in K^*$ such that $ab - \alpha ba \in sp(U) \subset sp(U_1)$. Clearly $sp(U_1)$ has greater dimension than $sp(U)$, thus in order to obtain a contradiction, it suffices to show that U_1 acts nilpotently on V.

Let m and r be integers such that $U^m = c^r = 0$ and let y be the product of any mr elements of U_1. If c occurs at least $mr - m + 1$ times in y, then at some point there must be at least r consecutive occurrences of c, hence $y = 0$. On the other hand, if c occurs at most $mr - m$ times then there will be at least m appearances of elements of U. Every time a term of the form cu appears in a product, where $u \in U$, it can be replaced by $\alpha(c, u)uc + u'$, where $u' \in sp(U)$. By repeatedly applying this procedure, we can shift elements of U to the left and y can be replaced by a linear combinations of products containing at least m consecutive elements of U on the left. Once again we see that $y = 0$, hence it follows that $U_1^{mr} = 0$ and so, U_1 acts nilpotently on V. $\square$

From Proposition 1, we now easily obtain our main result on modules over Lie color algebras. Note that although the Lie color algebra L is finite-dimensional, the L-module V need not be finite-dimensional.

THEOREM 2. *Let L be a finite-dimensional Lie color algebra and V an L-module. If every homogeneous $x \in L$ acts nilpotently on V, then L acts nilpotently on V. That is, there is an integer N such that the composition of the action of any N elements of L is 0 on V.*

Proof. Given the homomorphism $\psi : L \longrightarrow End_K(V)$, apply Proposition 1 with $A = \psi(L)$

and $S = \psi(\bigcup_{g \in G} L_g)$. Note that S satisfies the hypotheses of the proposition, since for all $x \in L_g$ and $y \in L_h$, $\psi(x)\psi(y) - \epsilon(g,h)\psi(y)\psi(x) = \psi([x,y]) \in S$. $\square$

We can now use Theorem 2 to prove the Lie color algebra analog of Engel's theorem.

COROLLARY 3. *If L is a finite-dimensional Lie color algebra such that ad_x is nilpotent for every homogeneous $x \in L$, then L is nilpotent.*

Proof. Consider the map $\psi : L \longrightarrow End_K(L)$ defined as $\psi(l) = \mathrm{ad}_l$, for all $l \in L$. The identity $[[x,y],z] = [x,[y,z]] - \epsilon(g,h)[y,[x,z]]$ for all $x \in L_g$, $y \in L_h$, and $z \in L$ implies that $\psi([x,y]) = \psi(x)\psi(y) - \epsilon(g,h)\psi(y)\psi(x)$. Thus L is an L-module and, by Theorem 2, there is an integer N such that the composition of the action of any N elements of L is 0 on L. Therefore $[x_N, [x_{N-1}, [\cdots, [x_1, x] \cdots]]] = 0$ for all $x, x_1, \ldots, x_N \in L$, thus L is nilpotent. $\square$

ACKNOWLEDGMENTS

The first author was supported by the University Research Council at DePaul University. Both authors were supported by KBN Grant 2 2012 91 02. Much of this work was done when the first author was a visitor at University of Warsaw, Białystok Division and he would like to thank the University for its hospitality.

REFERENCES

1. J. Bergen and S. Montgomery, *Smash products and outer derivations*, Israel J. of Math. **53** (1986), 321-345.

2. M. Cohen and S. Montgomery, *Group graded rings, smash products, and group actions*, Trans. A.M.S. **282** (1984), 237-258.

3. D. Fischman and S. Montgomery, *Biproducts of braided Hopf algebras and a double centralizer theorem for Lie color algebras*, to appear.

4. I. Kaplansky, *Lie Algebras and Locally Compact Groups*, Chicago Lectures in Mathematics, University of Chicago Press, Chicago, 1971.

5. V. K. Kharchenko, *Automorphisms and Derivations of Associative Rings*, Mathematics and its Applications, Soviet Series, vol. 69, Kluwer Academic Publishers, 1991.

6. S. Montgomery, *Fixed Rings of Finite Automorphism Groups of Associative Rings*, Lecture Notes in Mathematics, vol. 818, Springer-Verlag, 1980.

7. S. Montgomery, *Hopf Algebras and Their Actions on Rings*, Conference Board of the Mathematical Sciences, American Mathematical Society, Providence, Rhode Island, 1993.

Constructing Maximal Commutative Subalgebras of Matrix Rings

WILLIAM C. BROWN   Michigan State University, East Lansing, Michigan

1   Notation and History

Throughout this paper, k will denote an arbitrary field. By a k-algebra, we will mean an associative k-algebra with $1 \neq 0$. Algebra homomorphisms are assumed to take 1 to 1 and subalgebras are assumed to have the same 1 as the containing algebra. We will let $M_{p \times q}(k)$ denote the set of all $p \times q$ matrices with entries from k. When $p = q = n$, the k-algebra $M_{n \times n}(k)$ will be denoted by $M_n(k)$. We will assume throughout that $n \geq 2$.

We will let $\mathcal{M}_n(k)$ denote the set of all maximal, commutative k-subalgebras of $M_n(k)$. Thus, $R \in \mathcal{M}_n(k)$ if and only if R is a commutative k-subalgebra of $M_n(k)$ with the following property: If S is any commutative k-subalgebra of $M_n(k)$ such that $R \subseteq S$, then $R = S$. If $C(*)$ denotes the centralizer of $*$ in $M_n(k)$, then clearly a commutative k-subalgebra $R \subseteq M_n(k)$ is an element of $\mathcal{M}_n(k)$ if and only if $C(R) = R$. The basic problem concerning commutative subalgebras of $M_n(k)$ is to classify up to isomorphism all $R \in \mathcal{M}_n(k)$.

Suppose $R \in \mathcal{M}_n(k)$. Then $\dim_k(R) < n^2$. In particular, R is a (commutative) artinian ring. Thus, $R = R_1 \times \cdots \times R_p$ where each R_i is a local ring (i.e., a commutative ring containing precisely one maximal ideal). Let $V = M_{1 \times n}(k)$. Then V is a finitely generated, faithful (right) R-module for which $\operatorname{Hom}_R(V, V) \cong C(R) = R$ via the regular representation. As R-modules, $V \cong V R_1 \times \cdots \times V R_p$ and each $V R_i$ is a finitely generated, faithful R_i-module. The regular representation induces an isomorphism $R_i \cong \operatorname{Hom}_{R_i}(V_i, V_i)$ for each $i = 1, \ldots, p$. Thus, R_i can be identified with a maximal, commutative k-subalgebra of $M_{n_i}(k)$ where $n_i = \dim_k(V_i)$. Hence, each maximal, commutative k-subalgebra of $M_n(k)$ is a finite product of local algebras which are maximal, commutative k-subalgebras of possibly smaller matrix rings. For this reason, most papers which deal with the classification of algebras in $\mathcal{M}_n(k)$ assume that the algebra is local. In this note, we present two interesting constructions of local k-algebras in $\mathcal{M}_n(k)$ which have residue class field k.

We will use the notation $(*, **, k)$ to indicate that $*$ is a commutative k-algebra which is local with unique maximal ideal $**$ and residue class field k. If (R, J, k) is such an algebra and $R \in \mathcal{M}_n(k)$, then we will write $(R, J, k) \in \mathcal{M}_n(k)$. If (R, J, k) is a local k-subalgebra of $M_n(k)$, then J is nilpotent. Let $i(J)$ denote the index of nilpotency of J. Since $n \geq 2$, $(R, J, k) \in \mathcal{M}_n(k)$ implies $J \neq (0)$. Thus, $i(J) \geq 2$ whenever $(R, J, k) \in \mathcal{M}_n(k)$.

The study of local k-algebras in $\mathcal{M}_n(k)$ has a long history dating back at least to 1905 (See [6]). Here is a sampling of some of the more famous results about $\mathcal{M}_n(k)$.

(A) *If $k = \mathbb{C}$, the complex numbers, and $n \leq 6$, then there are only finitely many conjugacy classes of $(R, J, \mathbb{C}) \in \mathcal{M}_n(\mathbb{C})$.*

These classes are all known and completely described in [7]. Thus, if $(R, J, \mathbb{C}) \in \mathcal{M}_n(\mathbb{C})$ and $n \leq 6$, then R is conjugate (and hence isomorphic) to only one of finitely many algebras which are listed in [7]. In particular, the problem of classifying local algebras in $\mathcal{M}_n(\mathbb{C})$ is completely solved for all $n \leq 6$.

If $n \geq 7$, then we have quite a different sort of result.

(B) *If k is infinite and $n > 6$, then there are infinitely many pairwise nonisomorphic $(R, J, k) \in \mathcal{M}_n(k)$.*

In fact, there are infinitely many such (R, J, k) in (B) with $i(J) = 3$. Proofs of these assertions can also be found in [7]. At any rate, the problem of constructing local algebras in $\mathcal{M}_n(k)$ for large n is an interesting one which is still an active area of research.

A more specialized problem than classifying local algebras in $\mathcal{M}_n(k)$ is the study of their dimensions. This problem also has a rich history. The best general result to date is as follows:

(C) *If $R \in \mathcal{M}_n(k)$, then $(2n)^{2/3} - 1 < \dim_k(R) \leq [n^2/4] + 1$.*

In (C), $[n^2/4]$ denotes the greatest integer less than or equal to $n^2/4$. The upper bound in (C) was first proven by I. Schur for $k = \mathbb{C}$ in [6]. The general argument for any k was given by W. Gustafson in [4]. This upper bound is sharp as Example 1 below indicates. The lower bound in (C) was proven by T.J. Laffey in [5]. Laffey has also shown that $\dim_k(R) \geq [3n^{2/3} - 4]$ when $J(R)^3 = (0)$. Here $J(R)$ denotes the Jacobson radical of R. This bound is known to be sharp for infinitely many n.

A more detailed history of this subject can be found in [2]. We finish this short history with two examples which are relevant to the constructions in the next section.

Example 1: (Schur Algebras) Let $2 \leq n = r + s$ with r, s natural numbers such that $|r - s| \leq 1$. Let R be the following set of matrices in $M_n(k)$.

$$(1) \qquad R = \left\{ \left(\begin{array}{c|c} xI_r & Z \\ \hline O & xI_s \end{array} \right) \middle| \; x \in k, \; Z \in M_{r \times s}(k) \right\}.$$

Throughout this paper, I_α will denote the identity matrix of size $\alpha \times \alpha$. R is clearly a commutative k-subalgebra of $M_n(k)$ of dimension $\dim_k(R) = 1 + rs = [n^2/4] + 1$. R is local with Jacobson radical J consisting of those matrices in (1) with $x = 0$. Thus, $i(J) = 2$. It is easy to check that $(R, J, k) \in \mathcal{M}_n(k)$. ∎

The algebra given in Example 1 has dimension as large as possible by (C). In honor of Schur's early contributions to the results in (C), we call the algebra constructed in Example 1 a Schur algebra. Notice that any Schur algebra has dimension greater than or equal to n. For many years, it was conjectured that $(R, J, k) \in \mathcal{M}_n(k)$ implies $\dim_k(R) \geq n$. For example, if $i(J) = 2$, then $\dim_k(R) \geq n$. This was proven in [3]. In 1965, R.C. Courter gave the first example of a local $(R, J, k) \in \mathcal{M}_n(k)$ with $\dim_k(R) < n$. For our purposes, Courter's example can be described as follows:

Example 2: (Courter) Let $n = 14$. Set $R = J \oplus I_{14}k$ where J is the set of all 14×14 matrices of the following form:

$$(2) \qquad \begin{bmatrix} 0 & 0 & & & & & & & & & & & & \\ 0 & 0 & & & & O_{2\times10} & & & & & & O_{2\times2} & \\ \hline x_{11} & 0 & & & & & & & & & & & \\ 0 & x_{11} & & & & & & & & & & & \\ x_{12} & 0 & & & & & & & & & & & \\ 0 & x_{12} & & & & & & & & & & & \\ x_{21} & 0 & & & & O_{10\times10} & & & & & & O_{10\times2} & \\ 0 & x_{21} & & & & & & & & & & & \\ x_{22} & 0 & & & & & & & & & & & \\ 0 & x_{22} & & & & & & & & & & & \\ z_{11} & z_{12} & & & & & & & & & & & \\ z_{21} & z_{22} & & & & & & & & & & & \\ \hline y_{11} & y_{12} & z_{11} & z_{12} & z_{21} & z_{22} & 0 & 0 & 0 & 0 & x_{11} & x_{12} & 0 & 0 \\ y_{21} & y_{22} & 0 & 0 & 0 & 0 & z_{11} & z_{12} & z_{21} & z_{22} & x_{21} & x_{22} & 0 & 0 \end{bmatrix}$$

In (2), the x_{ij}'s, y_{ij}'s and z_{ij}'s are arbitrary elements from k. $O_{p \times q}$ denotes the zero matrix of size $p \times q$. J is a k-subspace of $M_{14}(k)$ which is closed under multiplication and consists of matrices which are nilpotent of index at most 3. Thus, (R, J, k) is a local, k-subalgebra of $M_{14}(k)$ with $\dim_k(R) = 13$ and $i(J) = 3$. In [3], Courter showed $(R, J, k) \in \mathcal{M}_{14}(k)$.   ■

It follows from results in [3] and [5] that Courter's example is the smallest example with respect to n or $i(J)$ for which $\dim_k(R) < n$. In the next section of this paper, we will give a general construction which gives Courter's Example and many other interesting examples with $\dim_k(R) < n$ as special cases.

2   The Constructions

If $R \in \mathcal{M}_n(k)$, then $V = M_{1 \times n}(k)$ is a finitely generated, faithful R-module for which $\operatorname{Hom}_R(V, V) \cong R$ via the regular representation. Conversely, suppose R is a finite dimensional, commutative k-algebra, V is a finitely generated, faithful R-module and $\operatorname{Hom}_R(V, V) \cong R$ via the regular representation. Then R is isomorphic to a maximal, commutative k-subalgebra of $M_n(k)$ where $n = \dim_k(V)$. Thus, constructing maximal, commutative k-subalgebras of $M_n(k)$ [for various n] is equivalent to constructing pairs (R, V) where R is a finite dimensional, commutative k-algebra, V is a finitely generated, faithful R-module and $\operatorname{Hom}_R(V, V) \cong R$ via the regular representation. In this section, we will give two different procedures for constructing such pairs.

Suppose B is a commutative ring and N is a B-module. Let r be a positive integer. We will let N^r denote the direct sum of r copies of N and $B \ltimes N^r$ the idealization of the B module N^r. Suppose $(b, n_1, \ldots, n_r)$ denotes a typical element of $B \ltimes N^r$. Thus, $b \in B$ and $n_1, \ldots, n_r \in N$. Then $B \ltimes N^r$ is a commutative ring with addition and multiplication defined in the usual ways:

$$(3) \qquad (b, n_1, \ldots, n_r) + (b', n_1', \ldots, n_r') = (b + b', n_1 + n_1', \ldots, n_r + n_r').$$

$$(b, n_1, \ldots, n_r)(b', n_1', \ldots, n_r') = (bb', n_1 b' + n_1' b, \ldots, n_r b' + n_r' b).$$

If B is a k-algebra, then N is a k-vector space via the embedding of k into B and $B \ltimes N^r$ is a k-algebra via $(b, n_1, \ldots, n_r)x = (bx, n_1 x, \ldots, n_r x)$ for $x \in k$.

Consider the B module $B^r \oplus N$. We will let $\langle b_1, \ldots, b_r, n \rangle$, $b_i \in B, n \in N$, denote a typical element in $B^r \oplus N$. It is easy to check that $B^r \oplus N$ is a $B \ltimes N^r$-module with scalar multiplication defined by

$$(4) \qquad \langle b_1, \ldots b_r, n \rangle (b, n_1, \ldots, n_r) = \langle b_1 b, \ldots, b_r b, nb + \sum_{i=1}^{r} n_i b_i \rangle.$$

If N is a finitely generated (faithful) B-module, then $B^r \oplus N$ is a finitely generated (faithful) $B \ltimes N^r$-module. The following theorem is proven in [2].

(D) *Suppose B is a commutative ring and N is a faithful B-module. For any positive integer r, $V = B^r \oplus N$ is a faithful $R = B \ltimes N^r$-module, and $\mathrm{Hom}_R(V,V) \cong R$ via the regular representation.*

If B is a finite dimensional k-algebra and N is a finitely generated B-module, then (D) implies $R = B \ltimes N^r \in \mathcal{M}_p(k)$ where $p = \dim_k(B^r \oplus N)$. If $B = (B,m,k)$ is a local k-algebra, then R is also local with Jacobson radical $J = m \ltimes N^r$ and residue class field k. Notice $i(J) = i(m) + 1$.

The construction in (D) can be used to produce local k-algebras $(R,J,k) \in \mathcal{M}_p(k)$ with $\dim_k(R) < p$. Suppose (B,m,k) is a finite dimensional, local k-algebra and N is a finitely generated, faithful B-module. Then $(R = B \ltimes N^r, J = m \ltimes N^r, k)$ is a local k-algebra in $\mathcal{M}_p(k)$ where $p = \dim_k(B^r \oplus N)$. Furthermore, $p - \dim_k(R) = (r-1)[\dim_k(B) - \dim_k(N)]$. Hence, if $r \geq 2$ and $\dim_k(B) > \dim_k(N)$, then $\dim_k(R) < p$. For example, suppose B is a Schur algebra given in Example 1. Then $(B,m,k) \in \mathcal{M}_n(k)$ with $i(m) = 2$ and $\dim_k(B) = [n^2/4] + 1$. Let $N = M_{1 \times n}(k)$. Then N is a finitely generated, faithful B-module and $\dim_k(B) > \dim_k(N)$ whenever $n \geq 4$. Thus, $(R = B \ltimes N^r, J = m \ltimes N^r, k) \in \mathcal{M}_p(k)$ where $p = r\dim_k(B) + n$. If $r \geq 2$ and $n \geq 4$, then $\dim_k(R) < p$. If $r = 2$ and $n = 4$, then the reader can easily check that $B \ltimes N^2$ is isomorphic to the algebra given in Example 2. Thus, Courter's Example can be constructed from a Schur algebra using (D).

The construction $(B,N) \to (R,V)$ given in (D) is called a C_1-construction in [1]. There is a second construction given in [1] which produces local algebras in $\mathcal{M}_n(k)$ which in general are not C_1-constructions. We briefly sketch this second construction.

Suppose (B,m,k) is a finite dimensional, local k-algebra. We assume $m \neq (0)$. Let N be a finitely generated, faithful B-module such that $\mathrm{Hom}_B(N,N) \cong B$ via the regular representation. We have seen that every local k-algebra $(B,m,k) \in \mathcal{M}_n(k)$ determines such a pair $(B, N = M_{1 \times n}(k))$. Let $\mathrm{Soc}(B)$ denote the socle of B. Then there exists a $z \in \mathrm{Soc}(B)$ such that $\dim_k(Nz) = 1$. A proof of this easy assertion can be found in [1]. Fix $z \in \mathrm{soc}(B)$ with $\dim_k(Nz) = 1$.

Let X be an indeterminate over B. Let B_1 and N_1 denote the following k-algebra and B-module respectively:

(5)
$$B_1 = \frac{B[X]}{(mX, X^2 - z)}, \quad N_1 = N \oplus Nz.$$

In Equation (5), $(mX, X^2 - z)$ denotes the ideal in $B[X]$ generated by mX and $X^2 - z$. Let x denote the image of X in B_1. Then $B_1 = B[x]$. Every element in B_1 can be written uniquely in the form $b + \alpha x$ for some $b \in B$ and $\alpha \in k$. B_1

is a finite dimensional, local k-algebra with Jacobson radical $m_{B_1} = m + kx$ and residue class field k. Also, $i(m_{B_1}) = \max\{3, i(m)\}$.

The B-module N_1 becomes a B_1-module when multiplication by x is defined as follows: $(n, n'z)x = (n'z, nz)$. The reader can check that this operation is well defined and that N_1 is a finitely generated, faithful B_1-module. The following theorem is proven in [1].

(E) *Suppose (B_1, N_1) is constructed from (B, N) as above. Then $\mathrm{Hom}_{B_1}(N_1, N_1) \cong B_1$ via the regular representation.*

The construction in (E) yields another method for producing local algebras in $\mathcal{M}_p(k)$. Suppose $(B, m, k) \in \mathcal{M}_n(k)$. Set $N = M_{1 \times n}(k)$. Choose $z \in \mathrm{Soc}(B)$ such that $\dim_k(Nz) = 1$ and construct the pair (B_1, N_1) given above. (E) implies $(B_1, m_{B_1}, k) \in \mathcal{M}_{n+1}(k)$. The construction $(B, N) \to (B_1, N_1)$ is called a C_2-construction. Example 11 in [2] is a C_2-construction, but not a C_1-construction. Courter's Example is a C_1-construction, but not a C_2-construction. Thus, the two constructions given here are independent of each other and provide new methods for producing maximal, commutative subalgebras of $M_n(k)$.

References

[1] W.C. Brown, "Two Constructions of Maximal Commutative Subalgebras of $n \times n$ Matrices", preprint.

[2] W.C. Brown and F.W. Call, "Maximal Commutative Subalgebras of $n \times n$ Matrices," Communications in Algebra, to appear.

[3] R.C. Courter, "The Dimension of Maximal Commutative Subalgebras of K_n," Duke Mathematical Journal, **32**, 225-232 (1965).

[4] W.H. Gustafson, "On Maximal Commutative Algebras of Linear Transformations," Journal of Algebra **42**, 557-563 (1976).

[5] T.J. Laffey, "The Minimal Dimension of Maximal Commutative Subalgebras of Full Matrix Algebras," Linear Algebra and its Applications, **71**, 199-212 (1985).

[6] I. Schur, "Zur Theorie der Vertauschbären Matrizen," J. Reine Angew. Math., **130**, 66-76 (1905).

[7] D.A. Suprunenko and R.I. Tyschkevich, "Commutative Matrices," Academic Press, New York, (1968).

Galois Extensions over Local Number Rings

LINDSAY N. CHILDS   State University of New York at Albany, Albany, New York

To Daniel Zelinsky, with best wishes.

Thirty years ago, generalizing classical Galois theory to extensions of commutative and non-commutative rings was a problem which attracted broad attention. Professor Zelinsky contributed several papers to the literature in this area. (In fact, my original thesis problem was to generalize a paper on Galois theory of continuous transformation rings by Rosenberg and Zelinsky [RZ55].) Thus it seems appropriate to offer a paper on Galois extensions to a proceedings in his honor.

The purpose of this paper is to show how formal groups can help give a complete description of the group of H-Galois extensions of the valuation ring R of a finite extension of Q_p, when H is a finite abelian p-power rank R-Hopf algebra with connected dual. Except for the section on embedding rank p group schemes in dimension one formal groups, the paper is an exposition of 20-year old results which should be more widely known than they seem to be.

My thanks to J. Lubin for permission to include his Theorem 3, and to Karl Zimmermann, David Moss and Alan Koch for many stimulating discussions.

41

1.   Basic definitions.

Let R be a commutative ring, S a commutative R-algebra which is a finitely generated projective R-module (= "S is a finite R-algebra").   Let G be a finite group of R-algebra automorphisms of S.   Then S is a Galois extension of R with group G if any of six well-known equivalent conditions (Theorem 1.3 of [CHR65]) hold. Of these, we single out:

c)   the R-module map j from the crossed product

$$D(S, G) = \{\sum_{\sigma \in G} s_\sigma u_\sigma \mid s_\sigma \text{ in } S\}$$

to $\text{End}_R(S)$ given by $j(s_\sigma u_\sigma)(t) = s_\sigma \sigma(t)$ for all t in S, is bijective;

e)   the R-module map

$$h: S \otimes_R S \to S \otimes_R \text{Hom}_R(RG, R) \cong \text{Hom}_S(SG, S)$$

given by $h(s \otimes t)(\sigma) = s\sigma(t)$ is an isomorphism;

f)   for any maximal ideal m of S and any $\sigma \neq 1$ in G, there is some s in S so that $\sigma(s) - s \notin m$.

When applied to an extension $S \supseteq R$ of local or global number rings, this last condition says that for any maximal ideal m of S, the inertia group $I_m = \{\sigma \text{ in } G \mid \sigma(s) \equiv s \pmod m\}$ is trivial. If I_m is trivial, then m is unramified over R.   Thus if S/R is Galois with group G, then S is unramified over R.   This means that if S and R are the rings of integers of a randomly chosen Galois extension $L \supseteq K$ of algebraic number fields with Galois group G, it is unlikely that S will be a Galois extension of R with group G (indeed, if S = Z, never); whereas if S ⊇ R are localizations at some prime of R of rings of integers of a Galois extension L ⊇ K of algebraic number fields, then with finitely many exceptions (= primes dividing the discriminant of S/R), S will be a Galois extension of R with group G.

Now let H be a cocommutative R-Hopf algebra which is finitely generated and projective as R-module. Let H^* = $Hom_R(H, R)$, a commutative R-Hopf algebra. Chase and Sweedler [CS69] extended conditions c) and e) above to finite R-algebras S such that S is an H-module algebra or an H^*-comodule algebra, as follows:

c)   the R-module map j: S ⊗ H → $End_R(S)$ given by j(s ⊗ h)(t) = sh(t) for all t in S, is bijective;

e)   if α: S → S ⊗ H^*, by α(t) = $\sum_{(t)} t_{(1)} \otimes t_{(2)}$, is the H^*-comodule structure map on S, then the R-module map

$$h: S \otimes_R S \to S \otimes_R H^*$$

given by h(s ⊗ t) = (s ⊗ 1)α(t) = $\sum_{(t)} st_{(1)} \otimes t_{(2)}$ is bijective.

If Y = Spec(S), X = Spec(R) and G = Spec(H^*) (which makes sense if H is cocommutative), then condition e) for a Galois extension is a translation of the statement that Y → X is a principal homogeneous space for the finite group scheme G. Thus while some of us were thinking (and continue to think) about Galois extensions, algebraic geometers (e.g. [MR70], [Ro73]) were

independently studying principal homogeneous spaces for group
schemes, as we will observe in more detail below.

Condition c) is the key to Galois descent. By Morita
theory, there is an equivalence of categories between the
category of left R-modules and the category of left $End_R(S)$-
modules, given by the functor $S \otimes_R -$. An S-module M therefore
descends, that is, is of the form $S \otimes_R M_0$ for some R-module M_0,
iff the S-action on M extends to an action by $End_R(S)$. Now if
S/R is a Hopf Galois extension with Hopf R-algebra H, then
getting an $End_R(S)$-action on M is equivalent to finding an
H-module action on M which is compatible with the S-module action
(i.e. so that one gets an action by the smash product $S \# H$).
This criterion for descent for Galois extensions is a conceptual
simplification of the general condition for faithfully flat
descent (c.f. [KO74] or Waterhouse [Wa79], Chapter 17). Of
course it also works for modules with structure such as algebras,
Hopf algebras, etc., provided the H-module action respects those
structures (e.g. is a measuring on algebras).

Condition f) for Galois extensions with group G does not
extend to Hopf Galois extensions in the sense of requiring S/R to
be unramified. In fact, if S is a Hopf Galois extension of R,
rings of integers of number fields, with Hopf algebra H, then the
discriminant of S/R = the discriminant of H^*/R. This observation
of Greither [Gr92] and others reduces to the unramified condition
when S is a Galois extension with group G, for then $H^* = RG^*$
$\cong R \oplus R \oplus \ldots \oplus R$ ($|G|$ copies of R) as R-algebra, hence the

discriminant of $H^* = R = \text{disc}(S/R)$, so no prime of R ramifies in

S. Since wild ramification is possible for Hopf Galois

extensions of number rings, Hopf Galois theory has interesting

potential applications in Galois module theory. See [Ch87],

[Gr92] or [Tay92].

2. Local Hopf algebras and formal groups.

The study of Galois extensions, even with abelian (=

commutative and cocommutative, and finite) Hopf algebras, over

valuation rings of local fields, is complicated by the richness

of the array of Hopf algebras over such rings. The rank p case

is understood by the work of Tate and Oort [TO70], but even

describing Hopf algebra orders inside KC, C cyclic of order p^2,

is difficult: see [Gr92] or [Un94]. Thus it seems to be useful

to approach the subject using formal groups. Such is the point

of view we will adopt in the remainder of the paper.

The starting point is

Oort's Embedding Theorem ([Oo67], [MR70], (5.1); [MZ70],

(2.4); [Oo74], (3.1)). Let K be a finite extension of $\mathbb{Q}_p$, with

valuation ring R. Let H be a finite abelian Hopf R-algebra which

is connected as an algebra and of p-power rank. Let N = Spec(H).

There is a connected p-divisible group scheme G over Spec(R) so

that N embeds in G.

It follows, e.g. by work of Lubin, that if N is a finite

subgroup of a formal group G, then there is a formal group G' and an isogeny $\bar{f}: G \to G'$, defined over R, with kernel N. (If G, G' are formal groups of dimension n, then an isogeny $f: G \to G'$ is an n-tuple of power series $\bar{f} = (f_1, \ldots, f_n)$ such that

$$\ker(\bar{f}) = \{ \bar{x} \in \bar{m}^n \mid \bar{f}(\bar{x}) = 0\}$$

is finite, where $\bar{m}$ is the maximal ideal of the valuation ring of the algebraic closure of K.) That is, N fits into a short exact sequence of formal groups,

$$0 \to N \to G \to G' \to 0 .$$

In this situation, N may be represented as N = Spec(H), where $H = R[[\bar{x}]]/(\bar{f})$ with comultiplication Δ induced by $\Delta(h(\bar{x})) = h(G(\bar{x}, \bar{y}))$. To see that H is a Hopf algebra, set

$$I = \langle f_1(\bar{x}), \ldots, f_n(\bar{x}) \rangle, \quad \text{and}$$
$$J = I \hat{\otimes} R[[\bar{y}]] + R[[\bar{x}]] \hat{\otimes} I \subseteq R[[\bar{x}]] \hat{\otimes} R[[y]] \cong R[[\bar{x}, \bar{y}]].$$

To show that Δ is well-defined, we observe that

$$\Delta(f_i(\bar{x})) = f_i(F_1(\bar{x}, \bar{y}), \ldots, F_n(\bar{x}, \bar{y})) ;$$

since $\bar{f}$ is a homomorphism from F to G, we have, for each i, that

$$f_i(F_1(\bar{x}, \bar{y}), \ldots, F_n(\bar{x}, \bar{y}))$$
$$= G_i(f_1(\bar{x}), \ldots, f_n(\bar{x}), f_1(\bar{y}), \ldots, f_n(\bar{y})) ,$$

from which it is clear that $\Delta(f_i(\bar{x}))$ is in J. The counit sends $\bar{x}$ to $\bar{0}$, and the antipode s, induced by

$$s(g(\bar{x})) = g([-1]_1(\bar{x}), \ldots, [-1]_n(\bar{x})) ,$$

is also well-defined modulo I because $[-1]_G \circ \bar{f} = \bar{f} \circ [-1]_F$. Thus H is in fact an R-Hopf algebra, and the set of S-valued points of Spec(H) is $\{\bar{a} \text{ in } S^n \mid f_i(\bar{a}) = 0 \text{ for } i = 1, \ldots, n\} = N(S)$. So N = Spec(H).

3. Principal homogeneous spaces for subgroups of formal groups

If N is a finite abelian group scheme defined over Spec(R), then PH(N) denotes the group of principal homogeneous spaces for N over Spec(R). This is the same as the group PH(H) of (isomorphism classes of) H-Galois objects over R, or, equivalently, the group Gal(H*) of H*-Galois extensions of R, where H* = Hom$_R$(H, R) is the dual Hopf algebra to H, representing the Cartier dual of N.

Over the years there has been some considerable interest in computing PH(N) in one or more of its equivalent formulations. To cite two recent examples: globally, there is a homomorphism, the Picard invariant map, from PH(N) to Cl(H*) given by viewing a Galois extension S of R with Hopf algebra H* as a rank one projective H*-module. Taylor [Tay92] has recently shown that in certain number-theoretic situations the kernel of the Picard invariant map relates closely to the values of certain p-adic L-functions. Greither's recent LNM [Gr92b] gives an exposition of work on Gal(H*) when H* = RG, G cyclic of prime power order, in both global and local settings.

When R is the valuation ring of a local field and H is a connected abelian R-Hopf algebra, then Oort's embedding theorem yields a rather nice description of PH(H). Namely, take the short exact sequence

$$0 \to N \xrightarrow{\;f\;} G \to G' \to 0$$

over $X = \mathrm{Spec}(R)$, and apply cohomology in the finite topology. Mazur did this in ([Mz70], Corollary 2.7), and showed that

$$PH(N) \cong H^0(X, G')/\mathrm{Im}(H^0(X, G)).$$

Here $H^0(X, G)$ is just the R-valued points of G. More explicitly, if G has dimension n, then $H^0(X, G) \cong m^n$, m the maximal ideal of R, with group operation on m^n defined by the formal group G: $\bar{\alpha} +_G \bar{\beta} = F(\bar{\alpha}, \bar{\beta})$. If we denote this group by $m_G^{\,n}$, then

Proposition 1. $PH(N) \cong m_{G'}^{\,n}/f(m_G^{\,n}).$

In this way, embedding N in a short exact sequence of formal groups gives an approach to understanding the Galois extensions of R with Hopf algebra H^*, $N = \mathrm{Spec}(H)$.

Utilizing this description of PH(N), Mazur and Roberts [MR70] determined the cardinality of PH(N), namely,

$$|PH(N)| = |N(R)| \cdot q^{\mathrm{ord}(\det(\mathrm{Jac}(f)(0)))}$$

where $N(R) = \mathrm{Alg}(H, R)$ is the group of R-points of N, $q = |R/m|$, and $\mathrm{Jac}(f)(0)$ is the Jacobian matrix of partial derivatives $\partial f_i/\partial x_j\big|_{\bar{x}=\bar{0}}$.

In the remainder of this paper we will look at these results when G and G' have dimension one.

4. Dimension one embeddings.

To utilize the description of PH(H) it is convenient to embed N = Spec(H) into a recognizable formal group. The Oort embedding generally embeds N in a formal group of large dimension (c.f. [OM], p. 333, Remark). In this section we look at dimension one embeddings. These are what one would hope to have if the Hopf algebra H is monogenic (i.e. a quotient of a polynomial ring in one variable): e.g. the Tate-Oort Hopf algebras.

Example 1. When R contains μ_p, the p th roots of unity, here is a way of embedding a generically split group scheme of rank p into a dimension one formal group.

Let ζ be a primitive pth root of unity in R and let $\lambda =$ $1 - \zeta$. Let $ab = \lambda$ in R where a and b are both non-units, and consider the Tate-Oort Hopf algebra $H_a = R[x]/(x^p - a^{p-1}x)$. Then $H_a \circledast K$ is split, since $x^p - a^{p-1}x$ has p roots in K, namely, $x = 0$ and $x = \omega^i a$ for ω a primitive p-1 st root of unity in K, $i = 0$, 1, ..., p-2.

Consider the formal group law $G_b(x, y) = x + y + bxy$. Then it is easy to verify by induction that the image of any natural number q under the map [ ]: $\mathbf{Z} \to End(G_b)$ is given by

$$[q](x) = \frac{(1+b)^q - 1}{b}.$$

In particular,

$$[p](x) = px + \binom{p}{2}bx^2 + \ldots + b^{p-1}x^p$$
$$= b^{p-1}(cx + \ldots + x^p)$$

for some c in R with $cR = a^{p-1}R$. Set

$$h(x) = [p](x)/b^{p-1} = cx + \ldots + x^p$$

Since $[p](x)$ is an endomorphism of G_b, it is easy to see that $h(x)$ is a homomorphism from G_b to G_{b^p} . Thus we have a short exact sequence of formal groups:

(1) $0 \to \mathbb{D} \to G_b \to G_{b^p} \to 0$

where $\mathbb{D} = \ker(h(x)) = \text{Spec}(R[x]/h(x))$.

Notice that $\mathbb{D}$ is generically split, since

$$h(x) = \frac{(1+b)^p - 1}{b^p}$$

has p roots in K since K contains μ_p. Thus $\mathbb{D}$ is determined by its discriminant, which is $h'(0)^p R = c^p R = (a^{p-1})^p R$. Thus $\mathbb{D} = \text{Spec}(H_a)$ and $R[x]/h(x) \cong H_a$.

From the sequence (1) and Proposition 1 above, we have a description of $PH(\mathbb{D})$, namely,

$$PH(\mathbb{D}) \cong G_{b^p}(R)/h(G_b(R))$$

where $G_b(R) = m$, the maximal ideal of R with group structure given by the formal group G_b On the other hand, Roberts [Ro73], and subsequently Hurley [Hu87] described $PH(\mathbb{D})$, namely,

$$PH(\mathbb{D}) \cong U_{b^p}(R)/U_b(R)^p$$

where $U_b(R) = \{u \in U(R) \mid u \equiv 1 \pmod{bR}\}$. (c.f. Greither [Gr92], II 2.1). We can recover Roberts' description from the formal group description, as follows:

Proposition 2. If p/b^p is in m, then the map

$$\psi: G_{b^p}(R)/h(G_b(R)) \to U_{b^p}(R)/U_b(R)^p$$

induced by sending s to $1 + b^p s$ for s in m, is an isomorphism of groups.

Proof. It is routine to check that the map sending s to $1 + b^p s$ is a homomorphism from $G_{b^p}(R)$ to $U_{b^p}(R)$ and induces a homomorphism on factor groups. To show ψ is 1-1, suppose for t in m, $\psi(t) = (1 + bs)^p = \psi(h(s))$. We must check that s is in m. But

$$t = \frac{(1+b)^p - 1}{b^p},$$

and upon expanding the right side, it becomes clear that, assuming that a is in m, then if t is in m, so is s.

To show ψ is onto, let $u = 1 + b^p s$ be in $U_{b^p}(R)$. If s is in m then u is in the image of ψ. If s is a unit, then let $-s \equiv t^p$ (mod m) and let $v = 1 + bt$. Then one checks easily that $uv^p = 1 + b^p w$ with w in m provided that p/b^p is in m, completing the proof.

This description of PH(D) is a formal group version of an argument of Greither [Gr92].

Example 2. Let F be a formal group of dimension one and height h defined over $R = O_K$, K a local number field. Suppose $N([p^n])$, the Newton polygon of $[p^n]_F(x)$, has a vertex at (p^r, b) where $r < nh$ and $b > 0$. Then there are homomorphisms $f: F \to F'$, $g: F' \to F$ of formal groups so that $[p^n] = g \circ f$ and $\ker(f)$ is a congruence subgroup of $\ker[p^n]$ of rank p^r consisting of all roots of $[p^n]$ whose valuations are the negatives of the slopes of the Newton polygon of $[p^n]$ to the left of (p^r, b). In this way we may represent $\ker(f)$ by $H = R[[x]]/(f(x))$. If we wish to realize H as a quotient of $R[x]$, we apply the Weierstrass Preparation Theorem to $f(x)$: $f(x) = h(x)u(x)$ where $h(x)$ is a Weierstrass polynomial, i.e. a monic polynomial of degree p^r which is congruent modulo m to x^{p^r}, whose Newton polygon is a vertical translate of that portion of the Newton polygon of $[p^n]$ with abscissas $\leq p^r$. Then $H \cong R[x]/h(x)$ as R-algebras.

One has considerable freedom in obtaining Hopf R-algebras in this way, by use of the standard generic dimension one formal group of height h. This is a formal group F_t defined over $Z_p[[t_1, \ldots t_{h-1}]]$ with the property that the endomorphism $[p]_t$ in $\operatorname{End}(F_t)$ has the form

$$(*) \qquad [p]_t(x) = pxg_0(x) + \sum_{i=1}^{h-1} t_i x^{p^i} g_i(x) + x^{p^h} g_h(x)$$

where $g_i(x)$ is a unit of $Z_p[[t_1, \ldots, t_i]][[x]]$ for all $i < h$, and $g_h(x)$ is a unit of $Z_p[[t_1, \ldots, t_{h-1}]][[x]]$. See, e.g., [Lu79]. Any specialization $t \to \alpha = (\alpha_1, \ldots, \alpha_{h-1})$ in R^{h-1}, R a complete normal local domain containing Z_p, gives a dimension one formal group F_α of height h defined over R.

Given any finite abelian p-group Γ of order p^r and exponent p^e, if K is a finite extension of $\mathbb{Q}_p$ with valuation ring R, which is sufficiently ramified over $\mathbb{Q}_p$, we may find for any h sufficiently large, a specialization α in R^{h-1} such that the Newton polygon of $[p^e]_\alpha(x)$ has a vertex with abscissa p^r, and the congruence subgroup of $\ker[p^e]$ corresponding to the vertex at p^r is isomorphic to Γ. For details see [CZ94].

For example, for fixed $b = \text{ord}(\beta)$, β in R such that $0 < b < 1 = \text{ord}(p)$, choose h so large that $b(p^h - 1) < p^h - p$. Let F_t be the standard generic formal group of height h. Set $t_1 = \beta$, and $t_2 = t_3 = \ldots = t_{h-1} = 0$. Then the Newton polygon of $[p]_\beta$ has a vertex at (p, b). For since $b(p^h - 1) < p^h - p$, we have

$$\frac{-1 + b}{p - 1} < \frac{-b}{p^h - p}$$

so the slope of the line joining $(1, 1)$ and (p, b) is less than the slope of the line joining (p, b) and $(p^h, 0)$. So by the Local Factorization Principle of Lubin [Lu79], $[p]_\beta = g \circ f$, where $f: F_\beta \to F'$ and $g: F' \to F_\beta$ for some formal group F', and $\ker f = \{\alpha \text{ in } \ker[p]_\beta \mid \text{ord}(\alpha) = \frac{1 - b}{p - 1}\} \cup \{0\}$ is a congruence subgroup of $\ker[p]_\beta$ consisting of all roots of $[p]_\beta$ of valuation $\geq \frac{1 - b}{p - 1}$. Then $H_f = R[[x]]/(f(x))$ is a rank p Hopf algebra which represents the group scheme $\ker(f)$.

By Lubin's Lemma (Lu64], Lemma 4.1.2), we can assume that $[p]_\beta$, f and g all are sums of terms of degree $\equiv 1 \pmod{p-1}$. So

$f(x) = cx + ux^p + \ldots$ where $\mathrm{ord}(c) = \mathrm{ord}(p/\beta$ and $\mathrm{ord}(u) = 0$.

By the Weierstrass Preparation Theorem, $f(x) = h(x)u(x)$ in $R[[x]]$

where $h(x) = c_0 x + x^p$ with $\mathrm{ord}(c_0) = \mathrm{ord}(c)$ and $u(x)$ is an

invertible power series.

   In this way, given any fixed discriminant $a^p R$, setting

$t_1 = \beta$ with $\mathrm{ord}(\beta) = p/a$, we can find a Tate-Oort Hopf algebra

$H_{c_0} = R[x]/(c_0 x - x^p)$ with that discriminant, as the representing

Hopf algebra of the kernel of some homomorphism of dimension one

formal groups.

   However, it is not clear from this construction which Tate-

Oort Hopf R-algebra H_{c_0} is (i.e what c_0 is), since varying c_0 by

a unit of R changes H_{c_0} generically (i.e. over K).

   By being more generic about the above construction, Lubin

has shown us

   Theorem 3.   Let R be the valuation ring of a finite

extension K of $\mathbb{Q}_p$. Any Tate-Oort R-Hopf algebra which is

connected with connected dual may be realized as the representing

algebra of a congruence subgroup of $[p]_F$ for some formal group F

of dimension one defined over R.

   Proof.   Let $H_b = R[x]/(x^p - bx)$ be a Tate-Oort Hopf R-

algebra, where $0 < \mathrm{ord}(b) < 1 = \mathrm{ord}(p)$. Suppose $\mathrm{ord}(b) > \dfrac{1}{r+1}$ for

some r, and choose h so large that $r(p-1) < p^h - p$. Let F_t be

the standard generic formal group of height h, as above, with

$[p]_t(x)$ as in (*), above. Specialize t_i to 0 for $i > 1$, and set $t_1 = z$ in $S = \mathbb{Z}_p[[z, p/z, p^r/z^{r+1}]]$, a complete local domain with maximal ideal $\mathfrak{M} = (z, p/z, p^r/z^{r+1})$. Lubin's Local Factorization Principle applies to specializations of this generality. Call the image of $[p]_t(x)$ by $[p]_z(x)$. Then $[p]_z(x) = px + zx^p u + \ldots$ with u a unit of S (use Lubin's Lemma, as needed), so the Newton polygon of $[p]_z(x)$ has a vertex with abscissa p if $\dfrac{p^{p^h - p}}{z^{p^h - p + (p-1)}}$ is in the maximal ideal $\mathfrak{M}$ of S. But

$$\frac{p^{p^h - p}}{z^{p^h - p + (p-1)}} = \frac{p^{p^h - p - r(p-1)}}{z^{p^h - p - r(p-1)}} \cdot \frac{p^{r(p-1)}}{z^{r(p-1) + (p-1)}}$$

is a product of elements in $\mathfrak{M}$, and so $[p]_z(x)$ has a vertex with abscissa p.

Then $[p]_z(x) = q_z(x) r_z(x)$ where $q_z(x)$ is a Weierstrass polynomial, $r_z(x)$ is a power series in $S[[x]]$, and under any specialization $z \to c$ of S to a discrete valuation ring, the valuation of any root of $q_c(x)$ is greater than the valuation of all roots of $r_c(x)$. Invoking Lubin's Lemma, we can assume $q_z(x) = (u(z)p/z)x + x^p$ where $u(z)$ is a unit of S.

Now we want to specialize z so that $u(z)p/z = b$, that is, so that $u(z)p - bz = 0$. Since $u(z)$ is a unit in S,

$$u(z) = \sum_{n=-\infty}^{\infty} a_n z^n$$

where each a_n is in $\mathbb{Z}_p$, a_0 is a unit of $\mathbb{Z}_p$ and for each $n > 0$, $\mathrm{ord}(a_{-n}) \geq nr/(r+1)$. (Here, $\mathrm{ord}(p) = 1$.) Thus the Newton polygon of $u(z)$ lies above the cone with vertex $(0, 0)$ defined by the half-lines from $(0, 0)$ through $(1, 0)$ and from $(0, 0)$ through $(-(r+1), r)$.

Suppose $\dfrac{1}{r+1} < \text{ord}(b) < 1$. Then the Newton polygon of $pu(z) - bz$ has vertices at $(0, 1)$ and at $(1, \text{ord}(b))$. For if $\text{ord}(b) > 1/(r+1)$, then $\dfrac{\text{ord}(b)-1}{1} > \dfrac{-r}{r+1}$, so the edge between $(0, 1)$ and $(1, \text{ord}(b))$ has a less negative slope than the edge out of $(0, 1)$ to the left; also, since $\text{ord}(b) < 1$, the edge out of $(1, \text{ord}(b))$ to the right has non-negative slope, hence there is a vertex at $(1, \text{ord}(b))$.

Now we apply a result of Lazard [Lz63]. Let $m = 1 - \text{ord}(b)$. Then the Laurent series $w(z) = pu(z) - bz$ converges for α in $\bar{K}$ of valuation m. To see this, we see easily that for $n > 0$, since $m > 0$, $\text{ord}(pa_n\alpha^n) \geq 1 + nm \to \infty$ as $n \to +\infty$, while for $n < 0$,

$$\text{ord}(pa_n\alpha^n) \geq 1 + (-n)\frac{r}{r+1} + n(1 - \text{ord } b)$$

$$= 1 - n((\text{ord } b - 1) + \frac{r}{r+1}).$$

which goes to $+\infty$ as n goes to $-\infty$ since $\text{ord}(b) - 1 > -\dfrac{r}{r+1}$.

Set $w(z) = \displaystyle\sum_{n=-\infty}^{\infty} c_n z^n$, where $c_i = pa_i$ for $i \neq 1$, $c_1 = b + pa_1$.

Following Lazard, for any m, set

$$\text{ord}(w, m) = \inf_i(\text{ord}(c_i) + im),$$

$$n(w, m) = \text{least } i \text{ so that } \text{ord}(w, m) = \text{ord}(c_i) + im,$$

$$N(w, m) = \text{largest } i \text{ so that } \text{ord}(w, m) = \text{ord}(c_i) + im.$$

Then for $m = 1 - \text{ord}(b) = $ the negative of the slope of the edge between $(0, 1)$ and $(1, \text{ord}(b))$ in the Newton polygon of $w(z)$, it is easy to check that $\text{ord}(w, m) = 1$, $n(w, m) = 0$, $N(w, m) = 1$.

Applying Proposition 2 of [Lz63], $w(z)$ factors in $K[x]$ into $w(z) = P(z)g(z)$ where $P(z)$ is a polynomial of degree $N(w, m) - n(w, m) = 1$ with a root of valuation m, and $g(z)$ is a Laurent series which converges for $z = \alpha$ of valuation m.

Since $P(z)$ has degree 1, there exists a root c of $w(z)$ in K of valuation $m = 1 - \text{ord}(b) > 0$. The map sending z to c is a local homomorphism from S to R, sending $\mathfrak{M}$ to m, because $\text{ord}(c) > 0$, $\text{ord}(p/c) = \text{ord}(b) > 0$ and

$$\text{ord}(p^r/c^{r+1}) > r - (r+1)\left(\frac{r}{r+1}\right) = 0$$

Specializing z to c yields

$$u(c)p/c = -b$$

and $[p]_c(x) = (-bx + x^p)v_c(x)$, where $v_c(x)$ is in $R[[x]]$. Thus the roots Γ of $-bx + x^p$ in $\bar{K}$ form a congruence subgroup of $\ker[p]_c$, and so there is a homomorphism $f_c: F_c \to G$ for some formal group G such that $\ker(f_c) = \Gamma$, and $H = R[[x]]/f_c(x)$ represents $\Gamma = \ker(f_c)$. By the Weierstrass preparation theorem,

$$f_c(x) = (-bx + x^p)h(x),$$

where $h(x)$ is an invertible power series in $R[[x]]$. Thus $H \cong R[x]/(-bx + x^p)$ as R-algebras. Since a Tate-Oort R-Hopf algebra is uniquely determined by its structure as R-algebra, the proof is complete.

The above is not exactly Lubin's argument: he showed that one could choose the height h of F_t to be 2; however the

Tate-Oort Hopf algebra need not represent a congruence subgroup
of [p] if the valuation of b is too small.

5.   Counting principal homogeneous spaces.

Let R be the valuation ring of K, a finite extension of $\mathbb{Q}_p$.
Let H = R[[x]]/(f) be a finite Hopf algebra which represents the
kernel of an isogeny f: F → G of formal groups of dimension one.
We've seen that then PHS(H) ≅ G(R)/f(F(R)) = $m/f(m)$.

In that setting, the local Euler characteristic theorem of
Mazur and Roberts ([MR70], Prop. 8.1) asserts that

(*) $|\text{PHS}(H)| = |\# \text{Alg}_R(H, R)| \cdot q^{\text{ord } f'(0)}$,

where q = |R/m| and ord is the valuation on K normalized so that
ord(p) = e, the absolute ramification index of K over $\mathbb{Q}_p$.

When F, G are formal groups of dimension one, then the proof
of Rasala ([MR70], pp. 225-6) may be obtained, using Proposition
1, as an application of ideas in Frohlich [Fr68]. Here is how
the proof goes.

Proposition 4. Suppose H = R[[x]]/(f(x)), where f:F → G is
a homomorphism of formal groups of dimension one. Suppose
$[p]_F(x) = f(x)u(x)$ where f(x) is a Weierstrass polynomial of
degree p^h and u(x) is a constant multiple of an invertible power
series. Suppose also the logarithm and exponential maps of F and
G converge in some neighborhood of zero. Then

$|\text{PHS}(H)| = |\# \text{Alg}_R(H, R)| \cdot q^{\text{ord } f'(0)}$.

Proof. Following Frohlich, given the formal group F of dimension one, let P(F) denote the group of points of F, namely, $P(F) = \bar{m}$, the maximal ideal of the valuation ring of the algebraic closure $\bar{K}$ of K, with group operation given by $+_F$. Let $P(F, K) = P(F) \cap K$. Set $\Lambda(F) = \bigcup \{\ker[p^n]_F: P(F) \to P(F)\}$, the torsion subgroup of P(F), and let $\Lambda(F, K) = \Lambda(F) \cap K$. Let ℓ_F in $\mathrm{Hom}(F_K, \mathbb{G}_{a,K})$ be the logarithm map. Then $\ker\{\ell_F: P(F) \to \bar{m}_+\} = \Lambda(F)$. If f is in $\mathrm{Hom}(F, G)$ then the diagram

$$
\begin{array}{ccc}
P(F) & \xrightarrow{\;\ell_F\;} & \bar{m}_+ \\
f\downarrow & & \downarrow f'(0) \\
P(G) & \xrightarrow{\;\ell_G\;} & \bar{m}_+
\end{array}
$$

commutes ([Fr68], p. 112). So we can build from ℓ_F and f the two pairs of exact sequences of K-valued points:

$$
\begin{array}{ccccccccc}
0 & \to & \Lambda(F,\ K) & \to & P(F,\ K) & \to & P(F,\ K)/\Lambda(F,\ K) & \to & 0 \\
 & & \downarrow f & & \downarrow f & & \downarrow & & \\
0 & \to & \Lambda(G,\ K) & \to & P(G,\ K) & \to & P(G,\ K)/\Lambda(G,\ K) & \to & 0
\end{array}
$$

and

$$
\begin{array}{ccccccccc}
0 & \to & P(F,\ K)/\Lambda(F,\ K) & \xrightarrow{\ell_F} & m_+ & \to & m_+/\ell_F(P(F,\ K)) & \to & 0 \\
 & & \downarrow f & & \downarrow f'(0) & & \downarrow & & \\
0 & \to & P(G,\ K)/\Lambda(G,\ K) & \xrightarrow[\ell_G]{} & m_+ & \to & m_+/\ell_G(P(G,\ K)) & \to & 0
\end{array}
$$

The snake lemma applied to the first pair of sequences gives

$$0 \to \ker f \to \ker f \to 0 \to \Lambda(G, K)/f(\Lambda(F, K)) \to$$

$$P(G, K)/f(P(F, K)) \to P(G, K)/\Lambda(G, K) \cdot f(P(F, K)) \to 0.$$

The snake lemma applied to the second pair of sequences gives

$$0 \to 0 \to T \to P(G, K)/\Lambda(G, K) \cdot f(P(F, K)) \to m_+/f'(0)m_+ \to U \to 0$$

where T, U are the kernel and cokernel of the map from
$m_+/\ell_F(P(F,K))$ to $m_+/\ell_G(P(G,K))$ induced by multiplication by
$f'(0)$.

Now whenever we have an exact sequence of finite groups, the
alternating product of the orders equals 1. The two sequences
above and the corresponding exact sequences of kernels and
cokernels then give:

(1) $|P(G, K)/f(P(F, K))|$

$= |\Lambda(G, K)/f(\Lambda(F, K))| \cdot |P(G, K)/\Lambda(G, K) \cdot f(P(F, K))|$

which describes $|P(G, K)/f(P(F, K))| = |PH(N)|$;

$|\ker f| \cdot |\Lambda(G, K)| = |\Lambda(F, K)| \cdot |\Lambda(G, K)/f(\Lambda(F, K))|$,

which describes the first factor in the right side of (1),
provided $\Lambda(G, K)$ and $\Lambda(F, K)$ are finite;

(2) $|T| \cdot |m_+/f'(0)m_+| = |P(G, K)/\Lambda(G, K) \cdot f(P(F, K))| \cdot |U|$

which describes the second factor in the right side of (1); and

$|T| \cdot |m_+/\ell_G(P(G,K))| = |m_+/\ell_F(P(F,K))| \cdot |U|$

which allows us to substitute in (2) for $|T| \cdot |U|^{-1}$. From these
we obtain

$$|P(G, K)/f(P(F, K))| = |\ker f| \cdot |m_+/f'(0)m_+| \cdot$$

$(|\Lambda(G, K)| \cdot |m_+/\ell_G(P(G,K))|^{-1}) \cdot (|\Lambda(F, K)| \cdot |m_+/\ell_F(P(F,K))|^{-1})^{-1}.$

Now $\ker f$ is the set of K-points of the group scheme
represented by $H = R[[x]]/f(x)$, so $\ker f \cong \mathrm{Alg}_R(H, R)$; and

$|m_+/f'(0)m_+| = q^{\text{ord}(f'(0))}$, as is easily seen. So to finish the proof it suffices to show that

$$|\Lambda(F, K)| = |m_+/\ell_F(P(F,K))|$$

and is finite, and similarly for G.

Now in the exact sequence

$$0 \to \Lambda(F, K) \to P(F, K) \overset{\ell_F}{\to} m_+ \to m_+/\ell_F(P(F,K)) \to 0$$

the middle groups are not finite. But for F of finite height, by Serre's theorem ([Fr68], p. 109), the logarithm map ℓ_F and its inverse e_F are both defined over K and define inverse valuation-preserving isomorphisms between $F(J_r)$ and J_r^+, where $J_r = \{\alpha \text{ in } R \mid \text{ord}(\alpha) > r\}$ for all r sufficiently large. The same is also true for $F = G_b$, since $\ell(x) = \dfrac{\log(1+bx)}{b}$ and $e(x) = \dfrac{e^{bx}-1}{b}$ and both $\log(1 + z)$ and e^z converge on J_r for r sufficiently large.

Since ℓ_F is valuation-preserving on J_r for r large, there is some r so that

$$\Lambda(F, K) \cap F(J_r) = \{0\}$$

Hence the map from $\Lambda(F, K)$ to $P(F, K)/F(J_r)$ is 1-1, and so ℓ_F induces an exact sequence

$$0 \to \Lambda(F, K) \to P(F, K)/F(J_r) \to m_+/J_r^+ \to m_+/\ell_F(P(F,K)) \to 0.$$

Now $P(F, K) = m = m_+$ and $F(J_r) = J_r = J_r^+$, as sets. Since $|m_+/J_r^+|$ is finite, $\Lambda(F, K)$ will be finite and

$$|\Lambda(F, K)| = |m_+/\ell_F(P(F,K))|$$

once we show that

$$|P(F, K)/F(J_r)| = |m_+/J_r^+| .$$

For each integer s, $0 \leq s \leq r$, $F(J_s)$, which $= J_s^+$ as sets, is a subgroup of $P(F,K)$, so we have a chain

$$P(F,K) = F(J_0) \supseteq F(J_1) \supseteq \ldots \supseteq F(J_r)$$

of subgroups; of course we also have the chain

$$m_+ = J_0^+ \supseteq J_1^+ \supseteq \ldots \supseteq J_r^+ .$$

But for each s, $0 < s \leq r$, the map ψ: $F(J_s) \to J_s^+/J_{s+1}^+$ defined by $\psi(a) = a + J_{s+1}^+ = \bar{a}$ yields an isomorphism

$$F(J_s)/F(J_{s+1}) \cong J_s^+/J_{s+1}^+ .$$

To see this, write

$$F(x, y) = x + y + xyu(x, y)$$

with $u(x, y)$ in $R[[x, y]]$. Then ψ is a homomorphism, for given a, b in $F(J_s)$, ab is in $F(J_{s+1})$, and so

$$
\begin{aligned}
\psi(\bar{a} +_F \bar{b}) &= \text{class of } a + b + abu(a, b) \\
&= \text{class of } a + b \quad \text{in } J_s^+/J_{s+1}^+ \\
&= \psi(\bar{a}) + \psi(\bar{b}).
\end{aligned}
$$

ψ is clearly onto, and $\ker(\psi) = \{a|\ \bar{a} = 0\} = F(J_{s+1})$.

Since $F(J_s)/F(J_{s+1}) \cong J_s^+/J_{s+1}^+$ for each s, $0 \leq s < r$, it follows quickly that the groups $P(F, K)/F(J_r)$ and m_+/J_r^+ have the same cardinality. Therefore $|\Lambda(F, K)| = |m_+/\ell_F(P(F,K))|$, completing the proof.

Corollary 5. If H_b is a Tate-Oort algebra, then

$$|PH(H_b)| = (\# \text{ of roots of } x^p - bx \text{ in } K) \cdot q^{\text{ord}(b)} .$$

Proof. By Theorem 3, H_b represents $\ker(f_c)$, where f_c is a homomorphism of formal groups which factors as

$f_c(x) = (-bx + x^p)v_c(x)$ for $v_c(x)$ a unit of $R[[x]]$. Then $f_c{}'(0) = -bv_c(0)$, and so $\mathrm{ord}(f_c{}'(0)) = \mathrm{ord}(b)$. The result follows immediately from Proposition 4.

References

[CHR65]   S. Chase, D. Harrison, A. Rosenberg, Galois theory and Galois cohomology of commutative rings, Memoirs Amer. Math. Soc. 52 (1965), 15-33.

[CS69]   S. Chase, M. Sweedler, Hopf Algebras and Galois Theory, Springer Lecture Notes in Math. 97, 1969.

[Ch87]   L. Childs, Taming wild extensions with Hopf algebras, Trans. Amer. Math. Soc. 304 (1987), 111-140.

[CZ94]   L. Childs, K. Zimmermann, Congruence-torsion subgroups of dimension one formal groups, J. Algebra (to appear).

[Fr68]   A. Frohlich, Formal Groups, Springer Lecture Notes in Math. 74 (1968).

[Gr92]   C. Greither, Extensions of finite group schemes, and Hopf Galois theory over a complete discrete valuation ring, Math. Z. 210 (1992), 37-67.

[Gr92b]   C. Greither, Cyclic Galois Extensions of Commutative Rings, Springer LNM 1534 (1992).

[Hu87]   S. Hurley, Galois objects with normal bases for free Hopf algebras of prime degree, J. Algebra 109 (1987), 292-318.

[KO74]   M.-A. Knus, M. Ojanguren, Theorie de la Descent et Algebres d'Azumaya, Springer LNM 389 (1974).

[Lz63]   M. Lazard, Les zeros des fonctions analytiques d'une variable sur un corps value complet, Publ. Math. IHES 14 (1963), p. 223-251.

[Lu64]   J. Lubin, One-parameter formal Lie groups over p-adic integer rings, Annals of Math. 80 (1964), 464-484.

[Lu79]   J. Lubin, Canonical subgroups of formal groups, Trans. Amer. Math. Soc. 251 (1979), 103-127.

[Mz70]   B. Mazur, Local flat duality, Amer. J. Math. 92 (1970), 343-361.

[MR70]   B. Mazur, L. Roberts, Local Euler characteristics, Inv. Math. 9 (1970), 201-234.

[OM68]   F. Oort, D. Mumford, Deformations and liftings of finite commutative group schemes, Inv. Math. 5 (1968), 317-334.

[Oo67]   F. Oort, Embeddings of finite group schemes into abelian schemes, mimeographed notes, Bowdoin College, 1967.

[Oo74]   F. Oort, Dieudonne modules of finite local group schemes,

[Ro73]   L. Roberts, The flat cohomology of group schemes of rank p, Amer. J. Math. 95 (1973), 688-702.

[RZ55]   A. Rosenberg, D. Zelinsky, Galois theory of continuous transformation rings, Trans. Amer. Math. Soc. 79 (1955), 429-452.

[TO70]   J. Tate, F. Oort, Group schemes of prime order, Ann. Sci. Ecole Norm. Sup. (4) 3 (1970), 1-21.

[Tay92]   M. Taylor, The Galois module structure of certain arithmetic principal homogeneous spaces, J. Algebra 153 (1992), 203-214.

[Un94]   R. Underwood, R-Hopf algebra orders in KC_{p^2}, J. Algebra, to appear.

[Wa79]   W. S. Waterhouse, Introduction to Affine Group Schemes, Springer, 1979.

University at Albany, State University of New York
Research partially supported by NSA grant # MDA90492H3025

Infinite Extensions of Simple Modules over Semisimple Lie Algebras

RANDALL P. DAHLBERG   Allegheny College, Meadville, Pennsylvania

1. INTRODUCTION

This paper uses the work of G. Hochschild and T. Levasseur to construct essential extensions of simple highest weight modules for semisimple Lie algebras [Hochschild, 1959, 1960, Levasseur, 1986]. The problem of constructing maximal essential extensions (injective hulls) of simple modules for semisimple Lie algebras is still unresolved.

In the case of a finite dimensional solvable Lie algebra $\mathfrak{b}$, the injective hull of the 1-dimensional trivial $\mathfrak{b}$ module, $E_{\mathfrak{b}}(\mathbb{C})$, is isomorphic to a sub-$U(\mathfrak{b})$-module algebra of the linear dual $U(\mathfrak{b})^*$ of the enveloping algebra $U(\mathfrak{b})$ [Levasseur, 1986, Theorem 2.2]. This $U(\mathfrak{b})$-module algebra is a polynomial algebra in $\dim_{\mathbb{C}} \mathfrak{b}$ variables and is a subalgebra of the algebra of representative functions $R(\mathfrak{b})$ of $U(\mathfrak{b})$ (the maximal locally finite submodule of $U(\mathfrak{b})^*$). Injective hulls for the other simple $\mathfrak{b}$-modules are obtained by applying automorphisms of $U(\mathfrak{b})^*$ to $E_{\mathfrak{b}}(\mathbb{C})$ and all can be explicitly realized as submodules of $R(\mathfrak{b})$ [Levasseur, 1986, Lemma 3.3]. In fact, for a finite dimensional simple $\mathfrak{b}$-module S, $E_{\mathfrak{b}}(S)$ is isomorphic to the tensor product (over the base field $\mathbb{C}$) of S with $E_{\mathfrak{b}}(\mathbb{C})$ [Dahlberg, 1984, Theorem 11]. Hence in the case of a solvable Lie algebra, the injective hulls of simple modules can be explicitly constructed relative to the choice of a Poincaré-Birkhoff-Witt basis of the enveloping algebra [Dahlberg, 1984, Levasseur, 1986].

In the case of a semisimple Lie algebra $\mathfrak{g}$, little is known about the structure of injective hulls of simple $\mathfrak{g}$-modules. It is an easy consequence of Weyl's Theorem on complete reducibility that essential extensions of simple $\mathfrak{g}$-modules cannot be locally finite [Dahlberg, 1989, Remark 1]. However, injective hulls of artinian $sl(2, \mathbb{C})$-modules are locally artinian [Dahlberg, 1989]. Also, if $\mathfrak{h}$ is a Cartan subalgebra of $\mathfrak{g}$, the $\mathfrak{h}$ locally finite dual of a Verma module $M(\lambda)$ for λ integral and dominant is an injective hull of $L(\lambda)$ in the category $\mathcal{O}$, but in general, this object is not injective in the category of all $\mathfrak{g}$-modules [Jantzen, 1983, 4.10].

In this paper we apply the results on injective hulls of simple modules over solvable Lie algebras to the standard Borel subalgebra $\mathfrak{b}^+$ of a simple Lie algebra $\mathfrak{g}$. We show that these modules are $\mathfrak{g}$-submodules of $U(\mathfrak{g})^*$ under the left translation action and are essential extensions of a submodule isomorphic to a simple $\mathfrak{g}$-module (Theorems 5.1 and 5.2). In particular, if $\mathfrak{b}$ is a Borel subalgebra of $\mathfrak{g}$ then the injective hull of the 1-dimensional trivial $\mathfrak{b}$-module, $E_{\mathfrak{b}}(\mathbb{C})$, may be realized as a subalgebra of $U(\mathfrak{g})^*$ which is stable under the left translation action of $\mathfrak{g}$, and thereby becomes a sub-$U(\mathfrak{g})$-module algebra of $U(\mathfrak{g})^*$. This sub-$U(\mathfrak{g})$-module algebra is an essential extension of the 1-dimensional trivial $\mathfrak{g}$-module $\mathbb{C}$ (Theorem 6). Other essential extensions can be obtained from this one by twisting the $\mathfrak{g}$-action via automorphisms of $U(\mathfrak{g})^*$ induced by the action of the adjoint group of $\mathfrak{g}$ on $U(\mathfrak{g})$. Since the resulting extensions are all submodules of $U(\mathfrak{g})^*$, we may sum these submodules to obtain an essential extension of the trivial module containing all of these submodules.

In the case of $\mathfrak{g} = sl(2, \mathbb{C})$, we can obtain more precise information about our extensions. Indeed, we construct an essential extension of each simple highest weight module $L(\lambda)$ which properly contains ${}^t M(\lambda)$, the $\mathfrak{h}$-finite dual of the Verma module $M(\lambda)$, and is locally artinian with simple factors $L(\lambda)$ and $L(s_\alpha.\lambda)$ in the finite dimensional case and $L(\lambda)$ otherwise (Theorems 7.3 and 7.4). Finally, we compute the adjoint action of $SL(2, \mathbb{C})$ on $E_{\mathfrak{b}+}(\mathbb{C})$.

I would like to thank Andy Magid for his continued support and encouragement, and Marc D. Montalvo for his work on the *Mathematica* programs which allowed me to compute the adjoint action of $SL(2, \mathbb{C})$ on $U(sl(2, \mathbb{C}))^*$.

2. The Linear Dual of $U(\mathfrak{g})$

Let $\mathfrak{g}$ be an n-dimensional Lie algebra over the field of complex numbers $\mathbb{C}$. Then $U(\mathfrak{g})$, the universal enveloping algebra of $\mathfrak{g}$, is a cocommutative Hopf algebra with comultiplication Δ defined by $\Delta(x) = x \otimes 1 + 1 \otimes x$ for $x \in \mathfrak{g}$, counit ϵ where ϵ is the augmentation map of $U(\mathfrak{g})$ with kernel $\mathfrak{g}U(\mathfrak{g}) = U(\mathfrak{g})\mathfrak{g} = I(\mathfrak{g})$, and antipode ν where ν is the principal antiautomorphism of $U(\mathfrak{g})$. The linear dual of $U(\mathfrak{g})$, $U(\mathfrak{g})^* = \hom_{\mathbb{C}}(U(\mathfrak{g}), \mathbb{C})$, is therefore a commutative and associative algebra over $\mathbb{C}$ with $< f_1 f_2, u > = < f_1 \otimes f_2, \Delta(u) >$ for $f_1, f_2 \in U(\mathfrak{g})^*, u \in U(\mathfrak{g})$ where we identify $\mathbb{C} \otimes \mathbb{C}$ with $\mathbb{C}$ and $< f, u > = f(u)$ for $f \in U(\mathfrak{g})^*, u \in U(\mathfrak{g})$. The multiplicative identity is ϵ (for details, consult [Sweedler, 1969] or [Dixmier, 1977, Chapter 2]). $U(\mathfrak{g})^*$ has a $U(\mathfrak{g})$-bimodule structure defined by $< afb, u > = < f, bua >$ for $a, b, u \in U(\mathfrak{g})$ and $f \in U(\mathfrak{g})^*$. Define $\gamma, \check{\rho} : \mathfrak{g} \longrightarrow \text{End } U(\mathfrak{g})^*$ by $\gamma(x)f = xf, x \in \mathfrak{g}, f \in U(\mathfrak{g})^*$ (left translation) and $\check{\rho}(x)f = f\check{x}$ (principal antiautomorphism followed by right translation). Then γ and $\check{\rho}$ are representations of $\mathfrak{g}$ by derivations on $U(\mathfrak{g})^*$ [Dixmier, 1977, 2.7.7]. These representations extended to $U(\mathfrak{g})$ make $U(\mathfrak{g})^*$ into a (left) $U(\mathfrak{g})$-module algebra [Sweedler, 1969, Example (c), p.154]. $U(\mathfrak{g})^*$ has an involution $f \mapsto \check{f}$ defined by $< \check{f}, u > = < f, \check{u} >$. The relationship between γ and ρ is $(\gamma(u)f)^{\vee} = \check{\rho}(u)\check{f}$. It is easy to see that $U(\mathfrak{g})^*$ is an injective cogenerator of the category of all left $U(\mathfrak{g})$-modules and hence contains an isomorphic copy of the injective hull, $E_{\mathfrak{g}}(S)$, of each simple left $U(\mathfrak{g})$-module. Fix an ordered basis $X_1 \ldots X_n$ of $\mathfrak{g}$ and define for $I = (i_1, \ldots, i_n), J = (j_1, \ldots, j_n) \in \mathbb{N}^n$

$$X^I = X_n^{i_n} \ldots X_1^{i_1}, \qquad |I| = \sum_{k=1}^n i_k,$$

$$I! = \prod_{k=1}^n i_k!, \qquad \delta_{IJ} = \prod_{k=1}^n \delta_{i_k, j_k}$$

where δ_{ij} denotes the Kronecker delta. Then $\{X^I/I!\}_{I \in \mathbb{N}^n}$ is a Poincaré-Birkhoff-Witt (PBW) basis for $U(\mathfrak{g})$, and with respect to this basis, we may identify the algebra $U(\mathfrak{g})^*$ with the algebra $\mathbb{C}[[x_1, \ldots, x_n]]$ where $x^I = x_1^{i_1} \ldots x_n^{i_n}$ is defined by

$$< x^I, X^J/J! >= \delta_{IJ}.$$

The algebra isomorphism is given by

$$U(\mathfrak{g})^* \longrightarrow \mathbb{C}[[x_1, \ldots, x_n]]$$
$$f \longmapsto \sum_I < f, X^I/I! > x^I \qquad \text{[Dixmier, 1977, 2.7.5]}.$$

We may now interpret γ and $\check{\rho}$ as representations of $\mathfrak{g}$ in the Lie algebra of derivations of $\mathbb{C}[[x_1, \ldots, x_n]]$ although we caution that these maps depend upon the choice of a PBW basis of $U(\mathfrak{g})$. Via either of these representations, $\mathbb{C}[[x_1, \ldots, x_n]]$ becomes a left $\mathfrak{g}$-module. If we let $U^{[q]}(\mathfrak{g})$ denote the linear span of the ordered monomials X^I for $|I| \geq q$ then $\mathbb{C}[x_1, \ldots, x_n]$ identifies with all $f \in U(\mathfrak{g})^*$ such that $f = 0$ on $U^{[q]}(\mathfrak{g})$ for $q \gg 0$. Another subalgebra of interest is the algebra of representative functions on $U(\mathfrak{g})$:

$$R(\mathfrak{g}) = \{ f \in U(\mathfrak{g})^* \mid \dim_{\mathbb{C}} fU(\mathfrak{g}) < \infty \}$$
$$= \{ f \in U(\mathfrak{g})^* \mid \dim_{\mathbb{C}} U(\mathfrak{g})f < \infty \} \quad \text{[Hochschild, 1959, p.99]}.$$

PROPOSITION. *Let* $\mathfrak{g} = \mathfrak{j} \oplus \mathfrak{k}$ *where* $\mathfrak{j}$ *and* $\mathfrak{k}$ *are subalgebras of* $\mathfrak{g}$ *and let*

$$\pi : U(\mathfrak{g}) = \mathfrak{j}U(\mathfrak{g}) \oplus U(\mathfrak{k}) \longrightarrow U(\mathfrak{k})$$

be the right $\mathfrak{k}$-*module projection. Then*

$$\pi^* : U(\mathfrak{k})^* \longrightarrow U(\mathfrak{g})^*$$
$$\text{where } < \pi^*(f), u >=< f, \pi(u) > \text{ for } f \in U(\mathfrak{k})^*, u \in U(\mathfrak{g})$$

is an algebra monomorphism. Furthermore, if

$$\gamma_{\mathfrak{k}} : \mathfrak{k} \longrightarrow Der_{\mathbb{C}} U(\mathfrak{k})^*$$
$$\text{and } \gamma_{\mathfrak{g}} : \mathfrak{g} \longrightarrow Der_{\mathbb{C}} U(\mathfrak{g})^*$$

denote the left translation representations of $\mathfrak{k}$ *and* $\mathfrak{g}$ *respectively, then*

$$\gamma_{\mathfrak{g}}(x)(\pi^*(f)) = \pi^*(\gamma_{\mathfrak{k}}(x)(f))$$
$$\text{for } x \in \mathfrak{k} \text{ and } f \in U(\mathfrak{k})^*.$$

Proof:. It is clear that $\mathfrak{j}U(\mathfrak{g})$ is a coideal of $U(\mathfrak{g})$. Hence π is a coalgebra epimorphism and therefore π^* is an algebra monomorphism [Sweedler, 1969, 1.4.7, 1.4.1]. The rest is an easy calculation. $\square$

Suppose now that we choose an ordered basis $X_1, \ldots, X_m$ for $\mathfrak{k}$ where $m = \dim \mathfrak{k}$. Then with respect to the corresponding PBW basis of $U(\mathfrak{k})$ we may identify $U(\mathfrak{k})^*$ with $\mathbb{C}[[y_1, \ldots, y_m]]$ where $< y_i, u >= 1$ if $u = X_i$ and is 0 otherwise. Extend this basis to an ordered basis $X_1, \ldots, X_n$ of $\mathfrak{g}$ and identify $U(\mathfrak{g})^*$ with $\mathbb{C}[[x_1, \ldots, x_n]]$ via the corresponding PBW basis of $U(\mathfrak{g})$ as above. Then $\pi^*(U(\mathfrak{k})^*)$ identifies with $\mathbb{C}[[x_1, \ldots, x_m]]$ where $\pi^*(y_i) = x_i$.

COROLLARY. *In the above notation, if for* $1 \leq i \leq m$

$$\gamma_{\mathfrak{k}}(X_i) = \sum_{j=1}^{m} q_{ij} \frac{\partial}{\partial y_j}, \qquad q_{ij} \in \mathbb{C}[[y_1, \ldots, y_m]]$$

then

$$\gamma_{\mathfrak{g}}(X_i) = \sum_{j=1}^{m} \pi^*(q_{ij}) \frac{\partial}{\partial x_j}.$$

By the Proposition and its Corollary we may now identify $U(\mathfrak{k})^*$ with $\pi^*(U(\mathfrak{k})^*)$ $\equiv \mathbb{C}[[x_1, \ldots, x_m]]$ as a sub-$U(\mathfrak{k})$-module algebra of $U(\mathfrak{g})^* \equiv \mathbb{C}[[x_1, \ldots, x_n]]$ under the left translation representation $\gamma \equiv \gamma_{\mathfrak{g}}$ restricted to the subalgebra $\mathfrak{k}$. However, since clearly

$$\pi^*(U(\mathfrak{k})^*) = \{f \in U(\mathfrak{g})^* \mid < f, jU(\mathfrak{g}) >= 0\}$$

it follows that $U(\mathfrak{k})^*$ is also a sub-$U(\mathfrak{g})$-module algebra of $U(\mathfrak{g})^*$ under γ [Dixmier, 1977, 2.7.16].

3. Injective Hulls of Simple $U(\mathfrak{b}^+)$-modules

Let $\mathfrak{b}$ be a solvable Lie algebra and set $\mathfrak{t} = [\mathfrak{b}, \mathfrak{b}]$. Then $\mathfrak{b}$ has a nilpotent subalgebra $\mathfrak{n}$ such that $\mathfrak{b} = \mathfrak{n} + \mathfrak{t}$ [Hochschild, 1960]. There is a basis [Levasseur, 1986] $X_1, \ldots, X_m$ of $\mathfrak{t}$ such that for $X \in \mathfrak{b}$

$$[X, \sum_{j \leq i} \mathbb{C}X_j] \subset \sum_{j \leq i} \mathbb{C}X_j \qquad \text{and}$$

$$[X_i, X_j] \in \sum_{p < \min(i,j)} \mathbb{C}X_p, \qquad 1 \leq i, j \leq m.$$

Complete this basis to a basis $X_1, \ldots, X_n$ of $\mathfrak{b}$ such that $X_{m+1}, \ldots, X_n$ is a basis of $\mathfrak{n}$ satisfying $\overline{X}_{m+1}, \ldots, \overline{X}_n$ is a basis of $\mathfrak{b}/\mathfrak{t} \cong \mathfrak{n}/\mathfrak{n} \cap \mathfrak{t}$. Note that

$$\mathfrak{b}_i = \sum_{j \leq i} \mathbb{C}X_j$$

is an ideal of $\mathfrak{b}$ for $1 \leq i \leq n$.

If V is a $U(\mathfrak{b})$-module and τ is an automorphism of $U(\mathfrak{b})$, we denote by $^{\tau}V$ the $U(\mathfrak{b})$-module which, as a vector space, is just V, and where the action of $U(\mathfrak{b})$ is given by $u.v = \tau(u).v$ for $u \in U(\mathfrak{b})$ and $v \in V$. Now let $\mathbb{C}_0$ denote the 1-dimensional trivial $\mathfrak{b}$-module. For $c = (c_1, \ldots, c_{n-m}) \in \mathbb{C}^{n-m}$ let τ_c denote the automorphism of $U(\mathfrak{b})$ defined by

$$\tau_c(X_i) = X_i \text{ for } 1 \leq i \leq m,$$
$$\tau_c(X_{m+i}) = X_{m+i} + c_i \text{ for } 1 \leq i \leq n - m.$$

Let I_c be the ideal of $U(\mathfrak{b})$ generated by $X_1, \ldots, X_m, X_{m+1} - c_1, \ldots, X_n - c_{n-m}$. Note that $I_c = \tau_c^{-1}(I(\mathfrak{b}))$ where $I(\mathfrak{b})$ is the augmentation ideal of $U(\mathfrak{b})$. Now $U(\mathfrak{b})/I_c = \mathbb{C}_c$ is a simple $U(\mathfrak{b})$-module, and every finite dimensional simple $U(\mathfrak{b})$-module is of this form. It is easy to see that $\mathbb{C}_c \cong {}^{\tau_c}\mathbb{C}_0$ and hence ${}^{\tau_c}E_\mathfrak{b}(\mathbb{C}_0) \cong E_\mathfrak{b}(\mathbb{C}_c)$.

Now let $\mathfrak{g} = \mathfrak{n}^- \oplus \mathfrak{h} \oplus \mathfrak{n}^+$ be the triangular decomposition of the semisimple Lie algebra $\mathfrak{g}$ relative to a choice of a Cartan subalgebra $\mathfrak{h}$ and a set of simple roots B of the root system R of $\mathfrak{g}$. The corresponding Borel subalgebra $\mathfrak{b}^+ = \mathfrak{h} \oplus \mathfrak{n}^+$ has a basis of the above form where $\mathfrak{t} = \mathfrak{n}^+$ and $\mathfrak{n} = \mathfrak{h}$. In fact, we can choose a Chevalley basis $X_\beta \, (\beta \in R)$, $H_\alpha \, (\alpha \in B)$ of $\mathfrak{g}$ where $[X_\alpha, X_{-\alpha}] = H_\alpha$, $[X_\alpha, X_\beta] = N_{\alpha,\beta} X_{\alpha+\beta}$, $N_{\alpha,\beta} \in \mathbb{Z}$ when $0 \neq \alpha + \beta \in R$ [Jantzen, 1983, 7.1]. Choose a basis $X_1 \equiv X_{\beta_1}, \ldots, X_r \equiv X_{\beta_r} \, (\beta_i \in R^+)$ of $\mathfrak{n}^+$ satisfying the above commutation properties and complete to a basis $X_1, \ldots, X_r, H_1 \equiv H_{\alpha_1}, \ldots, H_l \equiv H_{\alpha_l} \, (\alpha_i \in B)$ of $\mathfrak{b}^+$ where the H's above form a basis of $\mathfrak{h}$. From the form of the commutation relations, it is clear that the last l of the X_i's are of the form X_{α_i} where the α_i's are the simple roots. Applying the above to $\mathfrak{b} = \mathfrak{b}^+$ we have the following

THEOREM. *Identifying $U(\mathfrak{n}^+)^*$ with $\mathbb{C}[[x_1, \ldots, x_r]]$ and $U(\mathfrak{b}^+)^*$ with $\mathbb{C}[[x_1, \ldots, x_{r+l}]]$ relative to the PBW basis $H^B X^A / A! B!$ of $U(\mathfrak{b}^+)$ as above, we have*

(1) *$\gamma(X_i)$ and $\gamma(H_i)$ have expressions of the form*

$$\gamma(X_i) = \frac{\partial}{\partial x_i} + \sum_{j \leq p} p_{ij} \frac{\partial}{\partial x_j} \qquad \text{for } 1 \leq i \leq r$$

$$\gamma(H_i) = \frac{\partial}{\partial x_{r+i}} + \sum_{j \leq p} p_{r+i,j} \frac{\partial}{\partial x_j} \qquad \text{for } 1 \leq i \leq l$$

where $p = \min(i - 1, r)$ and $p_{ij} \in \mathbb{C}[x_1, \ldots, x_{j-1}]$.
(2) *$E_{\mathfrak{n}^+}(\mathbb{C}_0) \cong \mathbb{C}[x_1, \ldots, x_r]$.*
(3) *$E_{\mathfrak{b}^+}(\mathbb{C}_0) \cong \mathbb{C}[x_1, \ldots, x_{r+l}]$.*
(4) *$R(\mathfrak{b}^+) \cong \mathbb{C}[x_1, \ldots, x_{r+l}, e^{cx_j}; c \in \mathbb{C}, r + 1 \leq j \leq r + l]$.*
(5) *$E_{\mathfrak{b}^+}(\mathbb{C}_c) \cong \mathbb{C}[x_1, \ldots, x_{r+l}] \exp(\sum_{i=1}^l c_i x_{r+i})$ for $c = (c_1, \ldots, c_l) \in \mathbb{C}^l$.*

Proof. (1) follows from [Levasseur, 1986, 1.3]. (2) follows from [Dahlberg, 1984, Theorem 3] or [Levasseur, 1976, Theorem 3]. (3) follows from [Levasseur, 1986, 2.2]. (4) follows from [Hochschild, 1960]. Finally using (1) we have $\mathbb{C} \exp(\sum_{i=1}^l c_i x_{r+i}) \cong \mathbb{C}_c$ where $c = (c_1, \ldots, c_l)$. Hence by [Dahlberg, 1984, Theorem 11]

$$E_{\mathfrak{b}^+}(\mathbb{C}_c) \cong E_{\mathfrak{b}^+}(\mathbb{C}_0) \otimes \mathbb{C}_c \cong \mathbb{C}[x_1, \ldots, x_{r+l}] \otimes \mathbb{C} \exp(\sum_{i=1}^l c_i x_{r+i})$$

$$\equiv \mathbb{C}[x_1, \ldots, x_{r+l}] \exp(\sum_{i=1}^l c_i x_{r+i}).$$

□

4. A COMMUTATION PROPERTY OF $U(\mathfrak{n})$ FOR $\mathfrak{n}$ NILPOTENT

The following is based upon the work of M. Vergne [Vergne, 1970, p.88]. Let $\mathfrak{n}$ be a nilpotent Lie algebra with lower central series $\mathfrak{n} = \mathfrak{n}_1 \supseteq \mathfrak{n}_2 \supseteq \cdots \supseteq \mathfrak{n}_{d+1} = (0)$. The index d is called the index of nilpotency of $\mathfrak{n}$. This series is a finite filtration on $\mathfrak{n}$ consisting of the ideals $\mathfrak{n}_i = [\mathfrak{n}, \mathfrak{n}_{i-1}], i > 1$, which satisfy $[\mathfrak{n}_i, \mathfrak{n}_j] \subseteq \mathfrak{n}_{i+j}$. The graded vector space $gr(\mathfrak{n}) = \sum \mathfrak{n}_i/\mathfrak{n}_{i+1}$ has the stucture of a Lie algebra:

$$[X + \mathfrak{n}_{i+1}, Y + \mathfrak{n}_{j+1}] = [X, Y] + \mathfrak{n}_{i+j+1}.$$

Let $Y_1, \ldots, Y_n$ be a basis of $gr(\mathfrak{n})$ consisting of homogeneous elements and define w_i to be the degree of homogeneity of Y_i, i.e. $Y_i \in \mathfrak{n}_{w_i}/\mathfrak{n}_{w_i+1}$. Choose a representative X_i in $\mathfrak{n}$ for each Y_i and define the weight of X_i by $wt(X_i) = w_i$. Then $X_1, \ldots, X_n$ is a basis of $\mathfrak{n}$. For any finite sequence $M = \{i_1, \ldots, i_r\}$ from the set $\{1, \ldots, n\}$ define the length of M by $l(M) = r$, and define the weight of M by $wt(M) = w_{i_1} + \cdots + w_{i_r}$. We denote by M^* the sequence consisting of the same indices as M but rearranged in non-decreasing order. Finally, set $X_M = X_{i_1} \ldots X_{i_r}$ and call X_{M^*} an ordered monomial.

LEMMA [Vergne, 1970, Lemme 3, p. 88]. *Let M be a finite sequence of indices from the set $\{1, \ldots, n\}$. Then $X_M = X_{M^*} + \sum C_{A^*} X_{A^*}$ where $C_{A^*} \in \mathbb{C}, l(A^*) < l(M)$, and $wt(A^*) \geq wt(M)$.*

PROPOSITION. *Let $\mathfrak{n}$ be a nilpotent Lie algebra with index of nilpotency d, and let $I(\mathfrak{n})$ denote the augmentation ideal of $U(\mathfrak{n})$. Then for each positive integer m*

$$I(\mathfrak{n})^{dm} \subseteq U^{[m]}(\mathfrak{n}).$$

Proof. By hypothesis, $1 \leq wt(X_i) \leq d$ for $i = 1, \ldots, n$. Furthermore, if $wt(M) \geq dm$ then $X_M = X_{M^*} + \sum C_{A^*} X_{A^*}$ where $l(A^*) < l(M)$ and $wt(A^*) \geq wt(M)$ by the Lemma. Since $dl(A^*) \geq wt(A^*) \geq wt(M) \geq dm$, we have $l(A^*) \geq m$. Hence the right-hand side of the expression for X_M lies in $U^{[m]}(\mathfrak{n})$. Since $I(\mathfrak{n})^{dm}$ consists of linear combinations of monomials X_M with $wt(M) \geq l(M) \geq dm$, the result follows. $\square$

(Compare [Hochschild, 1960, Lemma 1])

5. ESSENTIAL EXTENSIONS OF SIMPLE MODULES

5.1. Let $\mathfrak{g} = \mathfrak{n}^- \oplus \mathfrak{h} \oplus \mathfrak{n}^+$ be the triangular decomposition of $\mathfrak{g}$ introduced in §3. Extend the basis $X_1, \ldots, X_r, H_1, \ldots, H_l$ of $\mathfrak{b}^+$ in §3 to a basis $X_1, \ldots, X_r, H_1,$ $\ldots, H_l, Y_r, \ldots, Y_1$ of $\mathfrak{g}$ where $X_{-\beta_i} \equiv Y_{\beta_i} \equiv Y_i \equiv {}^t X_i \equiv {}^t X_{\beta_i}$ where ${}^t Z$ denotes the image of $Z \in \mathfrak{g}$ under the Chevalley anti-automorphism. The Y_i's form a basis of $\mathfrak{n}^-$. Our corresponding PBW basis is

$$Y^C H^B X^A / A! B! C! =$$

$$Y_1^{c_1} \ldots Y_r^{c_r} H_l^{b_l} \ldots H_1^{b_1} X_r^{a_r} \ldots X_1^{a_1} / (a_1! \ldots a_r! b_1! \ldots b_l! c_1! \ldots c_r!)$$

We may identify $U(\mathfrak{b}^+)^*$ with the sub-$U(\mathfrak{g})$-module algebra $\mathbb{C}[[x_1, \ldots, x_{r+l}]]$ of $U(\mathfrak{g})^* \equiv \mathbb{C}[[x_1, \ldots, x_{2r+l}]]$.

THEOREM. *Let* $\mathfrak{g}$ *be a semi-simple Lie algebra. Let*

$$U(\mathfrak{b}^+)^* \xrightarrow{\pi^*} U(\mathfrak{g})^*$$

be the injective algebra homomorphism obtained from the projection

$$U(\mathfrak{g}) = \mathfrak{n}^- U(\mathfrak{g}) \oplus U(\mathfrak{b}^+) \xrightarrow{\pi} U(\mathfrak{b}^+)$$

of right $U(\mathfrak{b}^+)$-*modules. Then* $\pi^*(E_{\mathfrak{b}^+}(\mathbb{C}))$ *is a sub-*$U(\mathfrak{g})$-*module algebra of* $U(\mathfrak{g})^*$ *where* $\mathfrak{g}$ *acts by left translation. Furthermore,* $\pi^*(E_{\mathfrak{b}^+}(\mathbb{C}))$ *is an essential extension of the* 1-*dimensional trivial* $\mathfrak{g}$-*module* $\mathbb{C}$.

Proof. First, we note that $\pi^*(E_{\mathfrak{b}^+}(\mathbb{C})) = \mathbb{C}[x_1, \ldots, x_{r+l}]$. Since π^* is a homomorphism of left $\mathfrak{b}^+$-modules, it is clear that $\pi^*(E_{\mathfrak{b}^+}(\mathbb{C}))$ is a sub-$\mathfrak{b}^+$-module of $U(\mathfrak{g})^*$. From our choice of PBW basis and the Corollary of Proposition 2, it follows that $\gamma(Z)$ for $Z \in \mathfrak{b}^+ \subset \mathfrak{g}$ restricted to $\pi^*(U(\mathfrak{b}^+)^*)$ has the same form as $\gamma(Z)$ when thought of as a derivation of $U(\mathfrak{b}^+)^*$, i.e. the derivation has coefficients in $\mathbb{C}[x_1, \ldots, x_{r+l}]$. Therefore, it suffices to show that $\pi^*(E_{\mathfrak{b}^+}(\mathbb{C}))$ is stable under the derivations $\gamma(Y)$ for $Y \in \mathfrak{n}^-$. To do this, all that is needed is to show that the $\gamma(Y_i)\,|_{\pi^*(E_{\mathfrak{b}^+}(\mathbb{C}))}$ for $1 \leq i \leq r$ have coefficients in $\mathbb{C}[x_1, \ldots, x_{r+l}]$.

Let $\pi^* f \in \pi^*(E_{\mathfrak{b}^+}(\mathbb{C}))$. We must show that

$$\gamma(Y_i)\pi^* f \in \pi^*(E_{\mathfrak{b}^+}(\mathbb{C})) \text{ for all } \quad 1 \leq i \leq r$$

i.e. $\gamma(Y_i)\pi^* f = 0$ on $\mathfrak{n}^- U(\mathfrak{g})$ and $U^{[q]}(\mathfrak{b}^+)$ for $q \gg 0$. Now for some $m \in \mathbb{N}, f(U^{[m]}(\mathfrak{b}^+)) = 0$. Let us assume first that the root vector Y_i corresponds to a simple root $\alpha \in B$, i.e. $Y_i \equiv Y_\alpha \in \mathfrak{g}^{-\alpha} = \{x \in \mathfrak{g}\,|\,[h, x] = -\alpha(h)x \text{ for all } h \in \mathfrak{h}\}$ [Jantzen, 1983, 2.1]. We have

$$
\begin{aligned}
< \gamma(Y_\alpha)\pi^* f, Y^C H^B X^A > &= < \pi^* f, Y^C H^B X^A Y_\alpha > \\
&= < f, \pi(Y^C H^B X^A Y_\alpha > \\
&= < f, \pi(Y^C Y_\alpha H^B X^A + Y^C [H^B, Y_\alpha] X^A \\
&\quad + Y^C H^B [X^A, Y_\alpha]) > \\
&= < f, \pi(Y^C H^B [X^A, Y_\alpha]) >
\end{aligned}
$$

since $\pi(\mathfrak{n}^- U(\mathfrak{g})) = 0$ and $[H^B, Y_\alpha] \in Y_\alpha U(\mathfrak{h}) \subset \mathfrak{n}^- U(\mathfrak{g})$. Clearly, $\pi(Y^C H^B [X^A, Y_\alpha]) = 0$ for $|C| > 0$ so we may assume that $|C| = 0$. Now

$$H^B [X^A, Y_\alpha] = H^B \sum_{i=1}^{r} X_r^{a_r} \ldots [X_i^{a_i}, Y_\alpha] \ldots X_1^{a_1}.$$

Since $\alpha \in B$, there are only two cases: $[X_i, Y_\alpha] \in \mathfrak{n}^+$ or $[X_i, Y_\alpha] = H \in \mathfrak{h}$.
   Assume $[X_i, Y_\alpha] \in \mathfrak{n}^+$. Then

$$H^B X_r^{a_r} \ldots [X_i^{a_i}, Y_\alpha] \ldots X_1^{a_1} = H^B \sum_{j=0}^{a_i-1} X r^{a_r} \ldots X_i^{a_i-(j+1)}[X_i, Y_\alpha]X_i^j \ldots X_1^{a_1}$$

$$\in H^B I(\mathfrak{n}^+)^{|A|}.$$

Now for $|A| \geq dm$ where d is the index of nilpotency of $\mathfrak{n}^+$, we have $I(\mathfrak{n}^+)^{|A|} \subset U^{[m]}(\mathfrak{n}^+)$ by Proposition 4. Hence our term lies in $H^B I(\mathfrak{n}^+)^{|A|} \subset H^B U^{[m]}(\mathfrak{n}^+) \subset U^{[|B|+m]}(\mathfrak{b}^+)$, and thus f will vanish on this term for $|A|, |B| \gg 0$.

Assume next that $[X_i, Y_\alpha] = H \in \mathfrak{h}$. Then

$$H^B X_{\beta_r}^{a_r} \dots [X_{\beta_i}^{a_i}, Y_\alpha] \dots X_{\beta_1}^{a_1} = a_i H^B X_{\beta_r}^{a_r} \dots X_{\beta_{i+1}}^{a_{i+1}} H X_{\beta_i}^{a_i-1} X_{\beta_{i-1}}^{a_{i-1}} \dots X_{\beta_1}^{a_1}$$

$$- a_i(a_i - 1) H^B X_{\beta_r}^{a_r} \dots X_{\beta_{i+1}}^{a_{i+1}} X_{\beta_i}^{a_i-1} X_{\beta_{i-1}}^{a_{i-1}} \dots X_{\beta_1}^{a_1}$$

$$= a_i H^B H X_{\beta_r}^{a_r} \dots X_{\beta_{i+1}}^{a_{i+1}} X_{\beta_i}^{a_i-1} X_{\beta_{i-1}}^{a_{i-1}} \dots X_{\beta_1}^{a_1}$$

$$- \alpha(H) a_i H^B \sum_{j=i+1}^{r} X_{\beta_r}^{a_r} \dots X_{\beta_{i+1}}^{a_{i+1}} X_{\beta_i}^{a_i-1} X_{\beta_{i-1}}^{a_{i-1}} \dots X_{\beta_1}^{a_1}$$

$$- a_i(a_i - 1) H^B X_{\beta_r}^{a_r} \dots X_{\beta_{i+1}}^{a_{i+1}} X_{\beta_i}^{a_i-1} X_{\beta_{i-1}}^{a_{i-1}} \dots X_{\beta_1}^{a_1}.$$

The last expression belongs to $U^{[|B|+|A|-1]}(\mathfrak{b}^+)$ and again f will vanish on these terms for $|A|, |B| \gg 0$.

Hence we've shown that $\gamma(Y_\alpha)\pi^* f = 0$ on $\mathfrak{n}^- U(\mathfrak{g})$ and $U^{[q]}(\mathfrak{b}^+)$ for $q \gg 0$ so $\gamma(Y_\alpha)\pi^* f \in \mathbb{C}[x_1, \dots, x_{r+l}]$ and thus

$$\gamma(Y_\alpha)|_{\pi^*(E_{\mathfrak{b}+}(\mathbb{C}))} = \sum_{i=1}^{r+l} p_i \frac{\partial}{\partial x_i}$$

where the $p_i \in \mathbb{C}[x_1, \dots, x_{r+l}]$ and α is any simple root.

Now if $\beta \in R^+ \setminus B$ then $\beta = \alpha_1 + \dots + \alpha_k$ where $\alpha_i \in B$ and each partial sum $\alpha_1 + \alpha_2 + \dots + \alpha_i$ lies in R^+ [Humphreys, 1972, Corollary to Lemma 10.2A]. Hence

$$Y_\beta = [Y_{\alpha_k}, [\dots [Y_{\alpha_3}, [Y_{\alpha_2}, Y_{\alpha_1}]] \dots].$$

Since γ is a representation of $\mathfrak{g}$

$$\gamma(Y_\beta) = [\gamma(Y_{\alpha_k}), [\dots [\gamma(Y_{\alpha_3}), [\gamma(Y_{\alpha_2}), \gamma(Y_{\alpha_1})]] \dots].$$

By the above, each derivation $\gamma(Y_{\alpha_i})|_{\pi^*(E_{\mathfrak{b}+}(\mathbb{C}))}$ has coefficients in $\mathbb{C}[x_1, \dots, x_{r+l}]$. Hence so does $\gamma(Y_\beta)|_{\pi^*(E_{\mathfrak{b}+}(\mathbb{C}))}$. This implies that $\pi^*(E_{\mathfrak{b}+}(\mathbb{C}))$ is stable under the derivations $\gamma(Y_\beta)$ for $\beta \in R^+$, and hence has the structure of a sub-$U(\mathfrak{g})$-module algebra of $\mathbb{C}[[x_1, \dots, x_{2r+l}]] \equiv U(\mathfrak{g})^*$.

Finally, $\pi^*(E_{\mathfrak{b}+}(\mathbb{C}))$ is an essential extension of the trivial $\mathfrak{g}$-module $\mathbb{C}$. For if $0 \neq \pi^* f \in \pi^*(E_{\mathfrak{b}+}(\mathbb{C}))$, we can find $u \in U(\mathfrak{b}^+)$ such that $0 \neq c = uf \in \mathbb{C}$. But $\gamma(u)\pi^* f = \pi^*(uf) = \pi^*(c) = c$ is nonzero in $\mathbb{C}$. $\square$

COROLLARY. *In the notation of the Theorem,* $\pi^*(R(\mathfrak{b}^+))$ *is a sub-$U(\mathfrak{g})$-module algebra of $U(\mathfrak{g})^*$.*

Proof. From the proof of the Theorem,

$$\pi^*(R(\mathfrak{b}^+)) = \mathbb{C}[x_1, \dots, x_{r+l}; \exp(zx_{r+i}); x \in \mathbb{C}, 1 \leq i \leq l]$$

is stable under

$$\gamma(Y_\beta)|_{\mathbb{C}[x_1, \dots, x_{r+l}]} = \sum_{i=1}^{r+l} p_i \frac{\partial}{\partial x_i}, \qquad p_i \in \mathbb{C}[x_1, \dots, x_{r+l}] \quad \square$$

5.2. In order to construct essential extensions of the simple $\mathfrak{g}$-modules, we need more precise information about the form of the derivations $\gamma(Y_{\alpha_i}), 1 \leq i \leq l$. Recall our PBW basis from §5.1

$$Y^C H^B X^A / A! B! C! \text{ for } A, C \in \mathbb{N}^r \text{ and } B \in \mathbb{N}^l,$$

resulting from our ordered basis for $\mathfrak{g}$

$$X_1, \ldots, X_r, H_1, \ldots, H_l, Y_r, \ldots, Y_1.$$

The corresponding dual basis

$$x_1, \ldots, x_{2r+l} \in U(\mathfrak{g})^* \equiv \mathbb{C}[[x_1, \ldots, x_{2r+l}]]$$

is such that

$$x_i \text{ corresponds to } X_i \qquad 1 \leq i \leq r$$
$$x_{r+i} \text{ corresponds to } H_i \qquad 1 \leq i \leq l$$
$$x_{2r+l-i+1} \text{ corresponds to } Y_i \qquad 1 \leq i \leq r.$$

In particular, for a simple root $\alpha_i, 1 \leq i \leq l$,

$$x_{r-(l-i)} \text{ corresponds to } X_{\alpha_i} = X_{r-(l-i)}.$$

LEMMA. *With respect to our PBW basis in §5.1, we have for $1 \leq i \leq l$,*

$$\gamma(Y_{\alpha_i})|_{\pi^*(E_{\mathfrak{b}+}(\mathbb{C}))} = \sum_{j=1}^{r} p_{\alpha_i, j} \frac{\partial}{\partial x_j} + x_{r-(l-i)} \frac{\partial}{\partial x_{r+i}}$$

where $p_{\alpha_i, j} \in \mathbb{C}[x_1, \ldots, x_r]$, and

$$p_{\alpha_i, r-(l-i)} = -x_{r-(l-i)}^2.$$

Proof. The fact that the $p_{\alpha_i, j} \in \mathbb{C}[x_1, \ldots, x_{r+l}]$ is a consequence of Theorem 5.1. Now for $1 \leq j \leq r + l$, the coefficient of $\frac{\partial}{\partial x_j}$ is

$$\gamma(Y_{\alpha_i}) x_j = \sum_{I=(A,B,C) \in \mathbb{N}^{2r+l}} < x_j, Y^C H^B X^A Y_{\alpha_i} / A! B! C! > x^I.$$

Since $< x_j, \mathfrak{n}^- U(\mathfrak{g}) >= 0$, we may assume $C = \underline{0} \equiv (0, \ldots, 0) \in \mathbb{N}^r$. Now

$$(*) \qquad H^B X^A Y_{\alpha_i} = Y_{\alpha_i} H^B X^A + [H^B, Y_{\alpha_i}] X^A + H^B [X^A, Y_{\alpha_i}]$$

so x_j will vanish on the first term on the right-hand side and also on the second term since $[H^B, Y_{\alpha_i}] \in Y_{\alpha_i} U(\mathfrak{h})$. If $|B| > 0, x_j$ will vanish on the last term as well, so we only get a non-zero value when $I = (A, \underline{0}, \underline{0})$.

Case 1: $j = r + k, 1 \leq k \leq l$. Then $< x_{r+k}, u >= 1$ for $u = H_k$ and 0 otherwise. We can write

(**)
$$[X^A, Y_{\alpha_i}] = [X' X_{\alpha_i}^{a_{r-(l-i)}} X'', Y_{\alpha_i}]$$
$$= [X', Y_{\alpha_i}] X_{\alpha_i}^{a_{r-(l-i)}} X'' + X'[X_{\alpha_i}^{a_{r-(l-i)}}, Y_{\alpha_i}] X'' + X' X_{\alpha_i}^{a_{r-(l-i)}} [X'', Y_{\alpha_i}].$$

Now X' is an ordered monomial in the X_α's, $\alpha \neq \alpha_i$ simple, so $[X_\alpha, Y_{\alpha_i}] = 0$ implies $[X', Y_{\alpha_i}] = 0$. Also, $[X'', Y_{\alpha_i}] \in U(\mathfrak{n}^+)$ so $< x_{r+k}, X' X_{\alpha_i}^{a_{r-(l-i)}} [X'', Y_{\alpha_i}] >= 0$. Therefore,

$$< x_{r+k}, [X^A, Y_{\alpha_i}] > =< x_{r+k}, X'[X_{\alpha_i}^{a_{r-(l-i)}}, Y_{\alpha_i}] X'' >$$
$$=< x_{r+k}, X' a_{r-(l-i)} H_i X_{\alpha_i}^{a_{r-(l-i)}-1} X''$$
$$- X' a_{r-(l-i)}(a_{r-(l-i)} - 1) X_{\alpha_i}^{a_{r-(l-i)}-1} X'' >$$
$$= \begin{cases} 1, & \text{if } k = i \text{ and } A = (0\ldots,0, \underbrace{1}_{r-(l-i)}, 0, \ldots, 0) \\ \\ 0, & \text{otherwise.} \end{cases}$$

Thus the coefficient of $\frac{\partial}{\partial x_{r+k}}$ is $x_{r-(l-i)}$ when $k = i$ and is 0 otherwise.

Case 2: $j = r - (l - i)$. By reasoning similar to the above it follows from (*) that

$$p_{\alpha_i, r-(l-i)} = \gamma(Y_{\alpha_i}) x_{r-(l-i)}$$
$$= \sum_{I=(A,\underline{0},\underline{0}) \in \mathbb{N}^{2r+l}} < x_{r-(l-i)}, [X^A, Y_{\alpha_i}]/A! > x^I.$$

By (**) and the above discussion we have

$$[X^A, Y_{\alpha_i}] = X'[X_{\alpha_i}^{a_{r-(l-i)}}, Y_{\alpha_i}] X'' + X' X_{\alpha_i}^{a_{r-(l-i)}} [X'', Y_{\alpha_i}].$$

We claim that $[X'', Y_{\alpha_i}]$ cannot involve X_{α_i}, and hence $x_{r-(l-i)}$ will vanish on the last term in the sum on the right-hand side. But this is clear since $[X_\beta, Y_{\alpha_i}] \in \mathfrak{g}^{\beta - \alpha_i}$ and $\beta - \alpha_i \neq \alpha_i$ for any $\beta \in R^+$. Hence

$$< x_{r-(l-i)}, [X^A Y_{\alpha_i}] > =< x_{r-(l-i)}, X'[X_{\alpha_i}^{a_{r-(l-i)}}, Y_{\alpha_i}] X'' >$$
$$=< x_{r-(l-i)}, X' a_{r-(l-i)} H_i X_{\alpha_i}^{a_{r-(l-i)}-1} X''$$
$$- X' a_{r-(l-i)}(a_{r-(l-i)} - 1) X_{\alpha_i}^{a_{r-(l-i)}-1} X'' >$$
$$= \begin{cases} -2 & \text{for } A = (0, \ldots, 0, \underbrace{2}_{r-(l-i)}, 0, \ldots, 0) \\ \\ 0 & \text{otherwise.} \end{cases}$$

Thus $p_{\alpha_i, r-(l-i)} = -x_{r-(l-i)}^2$. $\square$

THEOREM. *Let* $(m_1, \ldots, m_l) \in \mathbb{C}^l$. *Then*

(1)

$$\gamma(U(\mathfrak{g})) \exp(\sum_{j=1}^{l} m_j x_{r+j}) \cong L(\lambda)$$

for $\lambda = \sum_{j=1}^{l} m_j \omega_j$ *where* $\omega_1, \ldots, \omega_l$ *are the fundamental weights.*

(2) $\mathbb{C}[x_1, \ldots, x_{r+l}] \exp(\sum_{j=1}^{l} m_j x_{r+j})$ *is an essential extension of*

$$\gamma(U(\mathfrak{g})) \exp(\sum_{j=1}^{l} m_j x_{r+j}) \cong L(\lambda).$$

Proof. Let $v = \exp(\sum_{j=1}^{l} m_j x_{r+j})$. By Theorem 3 (1),

$$\gamma(X_i)v = 0 \quad \text{for} \quad 1 \leq i \leq r$$
$$\text{and } \gamma(H_i)v = m_i v = \lambda(H_i)v \quad \text{for} \quad 1 \leq i \leq l.$$

Hence v is a maximal vector of highest weight λ in $U(\mathfrak{g})^*$. It follows from the Lemma that

$$\gamma(Y_{\alpha_i}^n / n!)v = \binom{m_i}{n} x_{r-(l-i)}^n v$$

and hence,

$$\gamma(Y_{\alpha_i}^{m_i+1})v = 0 \quad \text{for} \quad 1 \leq i \leq l$$

when $(m_1, \ldots, m_l) \in \mathbb{N}^l$. In this case, $\gamma(U(\mathfrak{g}))v \cong L(\lambda)$ is finite dimensional [Humphreys, 1972, Theorem 21.4]; otherwise, $\gamma(U(\mathfrak{g}))v \cong L(\lambda)$ is infinite dimensional. This proves (1). From the form of $\gamma(X_i)$ and $\gamma(H_i)$ given in Theorem 3 (1), it is easy to verify (2). $\square$

6. Essential Extensions of the Trivial Module

In this section we use the fact that $E_{\mathfrak{b}+}(\mathbb{C})$ can be identified with a $\mathfrak{g}$-submodule of $U(\mathfrak{g})^*$ under the left translation representation γ (Theorem 5.1) to produce other essential extensions of the trivial $\mathfrak{g}$-module $\mathbb{C}$. This is done by applying to $E_{\mathfrak{b}+}(\mathbb{C})$ automorphisms of $U(\mathfrak{g})^*$ induced by the action of the adjoint group on $U(\mathfrak{g})$. We may then sum these submodules of $U(\mathfrak{g})^*$ to produce an essential extension of $\mathbb{C}$. This procedure was suggested in [Levasseur, 1987] for the case $\mathfrak{g} = sl(2, \mathbb{C})$.

Let G be the adjoint group of the semisimple Lie algebra $\mathfrak{g}$. The action of G on $\mathfrak{g}$ may be extended naturally to an action via automorphisms on $U(\mathfrak{g})$. It is easy to see that these automorphisms are bialgebra automorphisms and hence define an action of G on $U(\mathfrak{g})^*$ by transport of structure, i.e.

$$< g.f, u > \equiv < f, g^{-1}.u > = < f, Ad(g)^{-1}(u) >$$
$$\text{where } g \in G, f \in U(\mathfrak{g})^*, \text{ and } u \in U(\mathfrak{g})$$

where Ad denotes the adjoint representation. This action has the following properties which are easily verified:

(1) $$g.(u.f) = (g.u)(g.f)$$

(2) $$g.\check{u} = \check{(g.u)}$$

for $g \in G, f \in U(\mathfrak{g})^*$, and $u \in U(\mathfrak{g})$

LEMMA. *Identify $E_{\mathfrak{b}+}(\mathbb{C})$ with the $\mathfrak{g}$-submodule $\pi^*(E_{\mathfrak{b}+}(\mathbb{C}))$ of $U(\mathfrak{g})^*$ as in Theorem 5.1. Then for $g \in G$, $g.E_{\mathfrak{b}+}(\mathbb{C})$ and $E_{g.\mathfrak{b}+}(\mathbb{C})$ are isomorphic as $U(g.\mathfrak{b}^+)$-modules.*

Proof. $U(\mathfrak{g})^*$ is an injective $U(g.\mathfrak{b}^+)$-module for each $g \in G$ [Dahlberg, 1984, Proposition 4]. Hence $U(\mathfrak{g})^*$ contains a submodule isomorphic to $E_{g.\mathfrak{b}+}(\mathbb{C})$, and we shall identify these two modules. Now $g.E_{\mathfrak{b}+}(\mathbb{C})$ is a $U(g.\mathfrak{b}^+)$-module by (1). Since G fixes $\mathbb{C}$, $g.E_{\mathfrak{b}+}(\mathbb{C})$ is an essential extension of the trivial $g.\mathfrak{b}^+$-module $\mathbb{C}$. Hence $g.E_{\mathfrak{b}+}(\mathbb{C}) \subset E_{g.\mathfrak{b}+}(\mathbb{C})$ and so $E_{\mathfrak{b}+}(\mathbb{C}) \subset g^{-1}.E_{g.\mathfrak{b}+}(\mathbb{C})$. But $g^{-1}.E_{g.\mathfrak{b}+}(\mathbb{C})$ is a $U(\mathfrak{b}^+)$-module via $u.(g^{-1}.f) = (g^{-1}g.u)(g^{-1}.f) = g^{-1}.((g.u).f) \in g^{-1}.E_{g.\mathfrak{b}+}(\mathbb{C})$ where $u \in U(\mathfrak{b}^+)$ and $f \in E_{g.\mathfrak{b}+}(\mathbb{C})$. Let $0 \neq g^{-1}.f \in g^{-1}.E_{g.\mathfrak{b}+}(\mathbb{C})$. There exists $g.u \in U(g.\mathfrak{b}^+)$ such that $0 \neq (g.u).f = c \in \mathbb{C}$. Hence $u.(g^{-1}.f) = g^{-1}.((g.u).f) = g^{-1}.c = c$, and so $g^{-1}.E_{g.\mathfrak{b}+}(\mathbb{C})$ is an essential extension of $\mathbb{C}$ as a $U(\mathfrak{b}^+)$-module. Therefore, $E_{\mathfrak{b}+}(\mathbb{C}) = g^{-1}.E_{g.\mathfrak{b}+}(\mathbb{C})$ and so $g.E_{\mathfrak{b}+}(\mathbb{C}) = E_{g.\mathfrak{b}+}(\mathbb{C})$. $\square$

THEOREM. *Let $\mathfrak{g}$ be a semisimple Lie algebra and let G denote the adjoint group of $\mathfrak{g}$.*

(1) *For any Borel subalgebra $\mathfrak{b}$ of $\mathfrak{g}$, $E_{\mathfrak{b}}(\mathbb{C})$ has a $\mathfrak{g}$-module structure.*
(2) *Identify $E_{\mathfrak{b}+}(\mathbb{C})$ with the $\mathfrak{g}$-submodule $\pi^*(E_{\mathfrak{b}+}(\mathbb{C}))$ of $U(\mathfrak{g})^*$ as in Theorem 5.1. Then*

$$E \equiv \sum_{g \in G} g.E_{\mathfrak{b}+}(\mathbb{C}) \cong \sum_{\text{Borel subalgebras } \mathfrak{b} \subset \mathfrak{g}} E_{\mathfrak{b}}(\mathbb{C})$$

is a $\mathfrak{g}$-submodule of $U(\mathfrak{g})^*$ which is an essential extension of the trival $\mathfrak{g}$-module $\mathbb{C}$.

Proof. (1) Any Borel subalgebra $\mathfrak{b}$ of $\mathfrak{g}$ can be written as $g.\mathfrak{b}^+$ for some $g \in G$ [Dixmier, 1977, 1.10.20]. By the Lemma , $E_{\mathfrak{b}}(\mathbb{C}) = E_{g.\mathfrak{b}+}(\mathbb{C}) = g.E_{\mathfrak{b}+}(\mathbb{C})$. Let $u \in U(\mathfrak{g})$ and $f \in E_{\mathfrak{b}+}(\mathbb{C})$. Then

$$u.(g.f) = (gg^{-1}.u).(g.f) = g.((g^{-1}.u).f)$$

by (1). By Theorem 5.1, $E_{\mathfrak{b}+}(\mathbb{C})$ may be considered as a $\mathfrak{g}$-submodule of $U(\mathfrak{g})^*$ and so $(g^{-1}.u).f \in E_{\mathfrak{b}+}(\mathbb{C})$. Thus $u.(g.f) \in E_{\mathfrak{b}}(\mathbb{C})$ and the result follows.

(2) By (1) each of the summands of E is a $\mathfrak{g}$-submodule of $U(\mathfrak{g})^*$ and hence the same is true for E. Suppose

$$0 \neq e = \sum_{i=1}^{n} g_i.f_i \in E$$

where $f_i \in E_{\mathfrak{b}+}(\mathbb{C})$ and $g_i \in G, g_i \neq g_j$ for $i \neq j$.

An easy induction argument on n yields a $u \in U(\mathfrak{g})$ such that $0 \neq u.e \in \mathbb{C}$ and hence E is an essential extension of $\mathbb{C}$ in $U(\mathfrak{g})^*$. $\square$

7. The Case $sl(2, \mathbb{C})$

Throughout this subsection, $\mathfrak{g}$ will denote sl $(2, \mathbb{C})$. Let X, H, Y denote the ordered Chevalley basis for $\mathfrak{g}$ where $[X, Y] = H, [H, X] = 2X$, and $[H, Y] = -2Y$. The ordered monomials

$$Y^{i_3} H^{i_2} X^{i_1} / (i_1! i_2! i_3!) \text{ for } I = (i_1, i_2, i_3) \in \mathbb{N}^3$$

form a PBW basis for $U(\mathfrak{g})$. We identify $U(\mathfrak{g})^*$ with $\mathbb{C}[[x_1, x_2, x_3]]$ with respect to this choice of PBW basis.

The following formulas will prove useful in the calculations below. For $n \in \mathbb{N}$:

(1) $\quad [Y, H^n] = \sum_{i=1}^{n} (-1)^{i-1} 2^i \binom{n}{i} Y H^{n-i} = \sum_{i=1}^{n} 2^i \binom{n}{i} H^{n-i} Y$

(2) $\quad [X, H^n] = \sum_{i=1}^{n} (-2)^i \binom{n}{i} H^{n-i} X$

(3) $\quad [H, X^n] = 2n X^n$

(4) $\quad [H, Y^n] = -2n Y^n$

(5) $\quad [Y, X^n] = -n X^{n-1} H - n(n-1) X^{n-1} = -n H X^{n-1} + n(n-1) X^{n-1}$

(6) $\quad [X, Y^n] = n Y^{n-1} H - n(n-1) Y^{n-1}$

PROPOSITION 7.1. *With respect to the ordered basis X, H, Y of $\mathfrak{g}$, the representation*

$$\gamma : \mathfrak{g} \longrightarrow Der_{\mathbb{C}} U(\mathfrak{g})^* \equiv Der_{\mathbb{C}} \mathbb{C}[[x_1, x_2, x_3]]$$

is given by

$$\gamma(X) = \frac{\partial}{\partial x_1}$$

$$\gamma(H) = -2x_1 \frac{\partial}{\partial x_1} + \frac{\partial}{\partial x_2}$$

$$\gamma(Y) = -x_1^2 \frac{\partial}{\partial x_1} + x_1 \frac{\partial}{\partial x_2} + e^{-2x_2} \frac{\partial}{\partial x_3}$$

and extending linearly to all of $\mathfrak{g}$.

Proof. This is an easy calculation using formulas $1, 3$, and 5 above. $\square$

Let $\mathfrak{h} = \mathbb{C}H$ denote the Cartan subalgebra of $\mathfrak{g}$ spanned by H, let $R = \{-\alpha, \alpha\}$ be the root system where $\alpha(H) = 2$, and let $\rho \in \mathfrak{h}^*$ be defined by $\rho = \frac{1}{2}\alpha$. We denote by $\mathfrak{b}^+$ the Borel subalgebra spanned by X and H.

PROPOSITION 7.2. *Let* $\lambda \in \mathfrak{h}^*, \lambda(H) = c \in \mathbb{C}$, *and let* $L(\lambda)$ *denote the simple* $\mathfrak{g}$-*module of highest weight* λ.

(1) *If* $c \in \mathbb{C} \setminus \mathbb{N}$ *then* $\gamma(U(\mathfrak{g}))e^{cx_2} = \mathbb{C}[x_1]e^{cx_2} \cong L(\lambda)$.

(2) *If* $c \in \mathbb{N}$ *then* $\gamma(U(\mathfrak{g}))e^{cx_2} = \sum_{i=0}^{c} \mathbb{C}x_1^i e^{cx_2} \cong L(\lambda)$.

(3) $\mathbb{C}[x_1, x_2]e^{cx_2}$ *is an essential extension of the simple* $\mathfrak{g}$-*module* $\gamma(U(\mathfrak{g}))e^{cx_2}$ *for all* $c \in \mathbb{C}$.

Proof. It is obvious from Proposition 7.1 that

$$\gamma(X)e^{cx_2} = 0$$
$$\text{and } \gamma(H)e^{cx_2} = ce^{cx_2} = \lambda(H)e^{cx_2}$$

so e^{cx_2} is a maximal vector of highest weight λ in $\mathbb{C}[[x_1, x_2, x_3]]$. An easy induction argument yields

$$\gamma(Y^n/n!)e^{cx_2} = \binom{c}{n} x_1^n e^{cx_2} \text{ for } n \in \mathbb{N}.$$

If $c \in \mathbb{C} \setminus \mathbb{N}$ then $\binom{c}{n} \neq 0$ for all $n \in \mathbb{N}$ and $\gamma(U(\mathfrak{g}))e^{cx_2} = \mathbb{C}[x_1]e^{cx_2}$. In this case it follows that $\mathbb{C}[x_1]e^{cx_2}$ is a torsion-free $U(\mathfrak{n}^-)$-module and so is isomorphic to $M(\lambda)$ which is simple [Dixmier, 1977, 7.1.8(vi), 7.6.24]. On the otherhand, if $c \in \mathbb{N}$ then $\binom{c}{c+1} = 0$ so $\gamma(Y^{c+1}/(c+1)!)e^{cx_2} = 0$ and $\gamma(U(\mathfrak{g}))e^{cx_2} = \sum_{i=0}^{c} \mathbb{C}x_1^i e^{cx_2}$ is a finite dimensional simple module [Dixmier, 1977, 7.2.4]. This proves (1) and (2).

It is clear from Proposition 7.1 that $\mathbb{C}[x_1, x_2]e^{cx_2}$ is a $\mathfrak{g}$-sub module of $\mathbb{C}[[x_1, x_2, x_3]]$. Suppose

$$p = \sum_{i=0}^{n} a_i x_1^i e^{cx_2} \in \mathbb{C}[x_1, x_2]e^{cx_2}$$

where $a_i \in \mathbb{C}[x_2], 0 \leq i \leq n$ and $a_n \neq 0$. Now $\gamma(X^n/n!)p = a_n e^{cx_2}$ so let

$$a_n = \sum_{j=0}^{m} b_j x_2^j$$

where

$$b_j \in \mathbb{C}, \ 0 \leq j \leq m, \text{ and } b_m \neq 0.$$

Then

$$\gamma((H - c)^m X^n/(m!n!))p = b_m e^{cx_2} \neq 0$$

and $\mathbb{C}[x_1, x_2]e^{cx_2}$ is an essential extension of $\gamma(U(\mathfrak{g}))e^{cx_2}$. $\square$

We mention that $\mathbb{C}[x_1, x_2]$ is not an injective $U(\mathfrak{g})$-module since the map $\phi : U(\mathfrak{g})Y \longrightarrow \mathbb{C}[x_1, x_2]$ defined by $\phi(Y) = x_2$ has no extension to $U(\mathfrak{g})$. For if Φ is an extension, then $x_2 = \gamma(Y)\Phi(1)$ for some $\Phi(1) \in \mathbb{C}[x_1, x_2]$. But then $x_2 = x_1 \left(-x_1 \frac{\partial \Phi(1)}{\partial x_1} + \frac{\partial \Phi(1)}{\partial x_2} \right)$ which is impossible. Hence $\mathbb{C}[x_1, x_2] \not\cong E_{\mathfrak{g}}(\mathbb{C})$.

Assume that $\lambda \in \mathfrak{h}^*$ is dominant and integral, i.e. $\lambda(H) = m \in \mathbb{N}$ and the corresponding simple $\mathfrak{g}$-module $L(\lambda)$ is finite dimensional. Consider the Verma module

$$M(\lambda) = U(\mathfrak{g}) \underset{U(\mathfrak{b}^+)}{\otimes} \mathbb{C}_\lambda;$$

here $\mathbb{C}_\lambda$ denotes the simple $\mathfrak{b}^+$-module where X acts trivially and H acts via multiplication by m. Set $v_0 = 1 \otimes 1$ and $v_{-1} = 0$. A basis for $M(\lambda)$ is

$$v_n = Y^n/n!v_0 = Y^n/n! \otimes 1.$$

The $\mathfrak{g}$-action is given by:

$$X.v_n = (m - (n-1))v_{n-1}$$
$$H.v_n = (m - 2n)v_n$$
$$Y.v_n = (n+1)v_{n+1} \text{ for } n \geq 0$$

as is easily seen using formulas 4 and 6 at the beginning of this subsection.

Consider next $M(\lambda)^*$, the linear dual of $M(\lambda)$, equipped with the following action:

$$Z.f(m) = f(^tZm) \text{ for } Z \in \mathfrak{g}, f \in M(\lambda)^*, m \in M(\lambda).$$

Here ^tZ denotes the image of Z under the Chevalley anti-automorphism. $M(\lambda)^*$ equipped with this action is usually written as $M(\lambda)^{(t)}$. Consider the $\mathfrak{g}$-submodule of $M(\lambda)^{(t)}$

$$^tM(\lambda) = \underset{\mu \in \mathfrak{h}^*}{\oplus} (M(\lambda)^{(t)})^\mu = \underset{\mu \in \mathfrak{h}^*}{\oplus} (M(\lambda)^\mu)^*;$$

it is easy to see that this module lies in the category $\mathcal{O}$ [Jantzen, 1983, 4.10]. We can give an explicit description of this module by defining f_n to be the linear functional dual to the basis element v_n of $M(\lambda)$ defined above, i.e. $f_n(v_k) = \delta_{nk}$. Then $^tM(\lambda) = \underset{n \geq 0}{\oplus} \mathbb{C}f_n$. Using the above formulas for the action of $\mathfrak{g}$ on $M(\lambda)$ we easily calculate the action of $\mathfrak{g}$ on $^tM(\lambda)$: for $n \geq 0$,

$$X.f_n = nf_{n-1}$$
$$H.f_n = (m - 2n)f_n$$
$$Y.f_n = (m - n)f_{n+1}.$$

We now have the following

PROPOSITION 7.3. *For $\lambda \in \mathfrak{h}^*$ integral and dominant we have*

$$\mathbb{C}[x_1]e^{mx_2} \cong {}^tM(\lambda).$$

Proof. The correspondence

$$x_1^n e^{mx_2} \mapsto f_n \text{ for } n \in \mathbb{N}$$

defines the isomorphism as is easily seen by Proposition 7.1. $\square$

THEOREM 7.4. *Let* $\lambda \in \mathfrak{h}^*$ *with* $\lambda(H) = c \in \mathbb{C}$.

(1) *Suppose* $c \in \mathbb{C} \setminus \mathbb{N}$. *Set* $W_m = 0$ *for* $m \in -\mathbb{Z}^+$ *and*

$$W_m = \sum_{k=0}^{m} \mathbb{C}[x_1] x_2^k e^{cx_2} \text{ for } m \in \mathbb{N}.$$

Then $(W_m)_{m \in \mathbb{Z}}$ *is a filtration of* $\mathbb{C}[x_1, x_2] e^{cx_2}$ *consisting of* $\mathfrak{g}$-*submodules where* $W_m/W_{m-1} \cong L(\lambda)$ *for all* $m \in \mathbb{N}$.

(2) *Suppose* $c \in \mathbb{N}$. *Set* $W_m = (0)$ *for* $m \in -\mathbb{Z}^+$, *and*

$$W_{2m} = \sum_{j=0}^{m-1} \mathbb{C}[x_1] x_2^j e^{cx_2} + \sum_{k=0}^{c} \mathbb{C} x_1^k x_2^m e^{cx_2},$$

$$W_{2m+1} = \sum_{j=0}^{m} \mathbb{C}[x_1] x_2^j e^{cx_2}$$

for $m \in \mathbb{N}$. *Then* $(W_m)_{m \in \mathbb{Z}}$ *is a filtration of* $\mathbb{C}[x_1, x_2] e^{cx_2}$ *by* $\mathfrak{g}$-*submodules where*

$$W_{2m}/W_{2m-1} \cong L(\lambda)$$
$$W_{2m+1}/W_{2m} \cong L(s_\alpha.\lambda)$$

for $m \in \mathbb{N}$.

Proof. (1) Clearly the W_m are $\mathfrak{g}$-submodules of $\mathbb{C}[x_1, x_2] e^{cx_2}$ in view of Proposition 7.1. Now

$$W_m/W_{m-1} = (\mathbb{C}[x_1] x_2^m e^{cx_2} + W_{m-1})/W_{m-1} \text{ for } m \geq 1$$

and has a basis given by $v_{n,m} = x_1^n x_2^m e^{cx_2} + W_{m-1}$ for $m \geq 0$. We have

$$\gamma(Y)(v_{n,m}) = (c-n)v_{n+1,m},$$
$$\gamma(H)(v_{n,m}) = (c-2n)v_{n,m},$$
$$\gamma(X)(v_{n,m}) = nv_{n-1,m}$$

so $\gamma(U(\mathfrak{g}))(v_{0,m}) = W_m/W_{m-1}$ and $v_{0,m}$ is a maximal vector of highest weight $c = \lambda(H)$. Since $c - n \neq 0$ for all $n \in \mathbb{Z}^+$ it follows that W_m/W_{m-1} is a torsion-free $U(\mathfrak{n}^-)$-module, and hence $W_m/W_{m-1} \cong M(\lambda) \cong L(\lambda)$ [Dixmier, 1977, 7.1.8, 7.6.24].

(2) A basis for

$$W_{2m}/W_{2m-1} = \sum_{j=0}^{m-1} \mathbb{C}[x_1] x_2^j e^{cx_2} + \sum_{k=0}^{c} \mathbb{C} x_1^k x_2^m e^{cx_2} \bigg/ \sum_{j=0}^{m-1} \mathbb{C}[x_1] x_2^j e^{cx_2}$$

is given by $v_{k,m} = x_1^k x_2^m e^{cx_2} + W_{2m-1}$ for $0 \leq k \leq c$. Since $\gamma(X)(v_{0,m}) = 0$ and $\gamma(H)(v_{0,m}) = cv_{0,m}, v_{0,m}$ is a maximal vector of highest weight $c = \lambda(H)$. In this case, $\gamma(Y^n/n!)(v_{k,m}) = \binom{c}{n} v_{n,m}$, so $W_{2m}/W_{2m-1} \cong L(\lambda)$.

Similarly, a basis for

$$W_{2m+1}/W_{2m} = \sum_{k \geq c+1} \mathbb{C}x_1^k x_2^m e^{cx_2} + W_{2m}/W_{2m}$$

is $w_{k,m} = x_1^{k+c+1} x_2^m e^{cx_2} + W_{2m}, k \geq 0$. We have for all $k \geq 0$,

$$\gamma(Y)(w_{k,m}) = -(k+1)w_{k+1,m}$$
$$\gamma(H)(w_{k,m}) = (-2(k+1) - c)w_{k,m}$$
$$\gamma(X)(w_{k,m}) = (k+c+1)w_{k-1,m}$$

so $\gamma(U(\mathfrak{g}))(w_{0,m}) = W_{2m+1}/W_{2m}$ and $w_{0,m}$ is a maximal vector of highest weight $-c - 2 = s_\alpha.\lambda(H)$. It is clear that W_{2m+1}/W_{2m} is a torsion-free $U(\mathfrak{n}^-)$−module and hence is isomorphic to $M(s_\alpha.\lambda) \cong L(s_\alpha.\lambda)$. $\square$

PROPOSITION 7.5. *With respect to the $\mathfrak{g}$-module structure of $U(\mathfrak{g})^* \equiv \mathbb{C}[[x_1, x_2, x_3]]$ under γ we have the following:*

(1) *For $n, k \in \mathbb{Z}^+$ we have*

$$\gamma(Y^k/k!)(x_3^n) = \sum_{i=1}^{n}(-1)^{k+i}\binom{k-1}{i-1}x_1^{k-i}x_3^{n-i}e^{-2ix_2}.$$

(2) $\gamma(U(\mathfrak{g}))(x_3^n) \cong M(0)$ *for $n \in \mathbb{Z}^+$.*

Proof. (1) This follows by induction on k. (2) Since $\gamma(X)(x_3^n) = 0 = \gamma(H)(x_3^n)$, x_3^n is a maximal vector of highest weight 0. Hence there is a surjective $\mathfrak{g}$-homomorphism $\phi : M(0) \longrightarrow \gamma(U(\mathfrak{g}))(x_3^n)$ defined by $\phi(1 \otimes 1) = x_3^n$ [Dixmier, 1977, 7.1.8(i)]. Now (1) implies that $\gamma(u)$ restricted to $\gamma(U(\mathfrak{g}))(x_3^n)$ is injective for all non-zero $u \in U(\mathfrak{n}^-)) = \mathbb{C}[Y]$, so ϕ is injective and the result follows [Dixmier, 1977, 7.1.8(vi)]. $\square$

Finally, let $G = SL(2, \mathbb{C})$ act on $U(\mathfrak{g})^* \equiv \mathbb{C}[[x_1, x_2, x_3]]$ as in §6. Since G acts via automorphisms (i.e. by substitution of variables), it suffices to calculate the translates of x_1, x_2, x_3 by an element $g \in G$. We shall see that it will be sufficient to calculate the translates of x_1, x_2, x_3 by elements of the form e^{tx} and e^{ty}. We have

$$X = e^{-tx}.x = x$$
$$H = e^{-tx}.h = h + 2tx$$
$$Y = e^{-tx}.y = y - th - t^2x.$$

Suppose that $Y^n = (y - th - t^2x)^n$, $n = 1, 2, 3, \ldots$ is expressed as a linear combination of ordered monomials $y^{i_3}h^{i_2}x^{i_1}$. The following results are easily proved by induction:

(1) The sum of those terms of the form y^k is

$$\sum_{k=1}^{n} \frac{(n-1)!}{(k-1)!}\binom{n}{k} t^{n-k}y^k.$$

(2) The sum of those terms of the form x^k is

$$\sum_{k=1}^{n} \frac{(n-1)!}{(k-1)!} \binom{n}{k} (-1)^k t^{n+k} x^k.$$

(3) The term involving only h is $-(n-1)! t^n h$.

Therefore, by using the above formulas, we obtain

(i) $Y^n \equiv n! t^{n-1} y - (n-1)! t^n h - n! t^{n+1} x, \quad n \geq 1$ (modulo higher degree terms)

and using (i) we can show

(ii) $\quad Y^i H^{n-i} \equiv -(-2)^{n-i} i! t^{i+1} x, \quad i \geq 0$ (modulo higher degree terms).

If $f_i \in U(\mathfrak{g})^*$ corresponds to $x_i \in \mathbb{C}[[x_1, x_2, x_3]]$ under the isomorphism $U(\mathfrak{g})^* \cong \mathbb{C}[[x_1, x_2, x_3]]$ then

$$e^{tx}.x_i = \sum_{(i_1, i_2, i_3) \in \mathbb{N}^3} < f_i, e^{-tx}.y^{i_3} h^{i_2} x^{i_1} > / i_1! i_2! i_3! x_1^{i_1} x_2^{i_2} x_3^{i_3}$$

$$= \sum_{(i_1, i_2, i_3) \in \mathbb{N}^3} < f_i, Y^{i_3} H^{i_2} X^{i_1} > / i_1! i_2! i_3! x_1^{i_1} x_2^{i_2} x_3^{i_3}$$

so it suffices to find the linear term of $Y^{i_3} H^{i_2} X^{i_1}$ when expressed as a linear combination of the ordered monomials $y^{\nu_3} h^{\nu_2} x^{\nu_1}$ since f_i vanishes on all other ordered monomials. Since $X = x$, it is clear that we only need to consider the expressions $Y^{i_3} H^{i_2}$. By routine calculations using (i) and (ii) above we obtain

$$e^{tx}.x_1 = x_1 + t - \frac{t e^{-2x_2}}{1 - t x_3};$$

$$e^{tx}.x_2 = x_2 + \log(1 - t x_3);$$

$$e^{tx}.x_3 = \frac{x_3}{1 - t x_3}.$$

Similarly, for the action of e^{-ty} we have

$$X = e^{-ty}.x = -t^2 y + th + x$$

$$H = e^{-ty}.h = h - 2ty$$

$$Y = e^{-ty}.y = y.$$

Suppose that $X^n = (-t^2 y + th + x)^n, \quad n = 1, 2, 3, \ldots$ is expressed as a linear combination of ordered monomials $y^{i_3} h^{i_2} x^{i_1}$. Again we have similar results as above:

(1) The sum of those terms of the form x^k is

$$\sum_{k=1}^{n} \frac{(n-1)!}{(k-1)!} \binom{n}{k} (-1)^{n-k} t^{n-k} x^k.$$

(2) The sum of those terms of the form y^k is

$$(-1)^n \sum_{k=1}^{n} \frac{(n-1)!}{(k-1)!} \binom{n}{k} t^{n+k} y^k.$$

(3) The term involving only h is $(-1)^{n-1}(n-1)! t^n h$.

Similarly, by using the above formulas, we obtain

(iii) $\qquad X^n \equiv (-1)^n(n!t^{n+1}y - (n-1)!t^n h - n!t^{n-1}x), \quad n \geq 1$

modulo higher degree terms, and using (iii) we can show

(iv) $\quad H^{n-i}X^i \equiv (-1)^i(-2)^{n-i}i!t^{i+1}y, \quad i \geq 0$ (modulo higher degree terms).

Using (iii) and (iv) one can calculate the action of e^{ty} on x_1, x_2, x_3:

$$\exp(ty) \cdot x_1 = \frac{x_1}{1 + tx_1}$$

$$\exp(ty) \cdot x_2 = x_2 + \log(1 + tx_1)$$

$$\exp(ty) \cdot x_3 = x_3 - t + \frac{te^{-2x_2}}{1 + tx_1}.$$

The maximal torus T of diagonal matrices $g = \text{diag}(u, u^{-1})$ in G acts via

$$g \cdot x_1 = u^{-2}x_1$$

$$g \cdot x_2 = x_2$$

$$g \cdot x_3 = u^2 x_3.$$

Now let

$$g = \begin{pmatrix} a & b \\ c & d \end{pmatrix} \in SL(2, \mathbb{C}).$$

There are two cases:

(i) $a \neq 0$. Then

$$g = \begin{pmatrix} 1 & 0 \\ a^{-1}c & 1 \end{pmatrix} \begin{pmatrix} 1 & ab \\ 0 & 1 \end{pmatrix} \begin{pmatrix} a & 0 \\ 0 & a^{-1} \end{pmatrix}$$

$$= \exp(a^{-1}cy)\exp(abx)\text{diag}(a, a^{-1}).$$

Thus we have

$$g \cdot x_1 = \frac{(b + dx_1)(d - bx_3) - bde^{-2x_2}}{(a + cx_1)(d - bx_3) - bce^{-2x_2}}$$

$$g \cdot x_2 = x_2 + \log[(a + cx_1)(d - bx_3) - bce^{-2x_2}]$$

$$g \cdot x_3 = \frac{(a + cx_1)(-c + ax_3) + ace^{-2x_2}}{(a + cx_1)(d - bx_3) - bce^{-2x_2}}$$

(ii) $a = 0$. Then

$$g = \begin{pmatrix} 0 & b \\ -b^{-1} & d \end{pmatrix}$$

$$= \begin{pmatrix} 0 & 1 \\ -1 & 0 \end{pmatrix} \begin{pmatrix} 1 & -b^{-1}d \\ 0 & 1 \end{pmatrix} \begin{pmatrix} b^{-1} & 0 \\ 0 & b \end{pmatrix}$$

$$= \exp(x)\exp(-y)\exp(x)\exp(-b^{-1}dx)\text{diag}(b^{-1}, b).$$

We have

$$g \cdot x_1 = -bd + \frac{b(d - bx_3)}{x_1 x_3 - b^{-1}dx_1 + e^{-2x_2}}$$

$$g \cdot x_2 = x_2 + \log(x_1 x_3 - b^{-1}dx_1 + e^{-2x_2})$$

$$g \cdot x_3 = \frac{-b^{-2}x_1}{x_1 x_3 - b^{-1}dx_1 + e^{-2x_2}}.$$

REFERENCES

1. R.P.Dahlberg, *Injective hulls of Lie modules*, J.Algebra **87** (1984), 458–471.
2. _____, *Injective hulls of simple sl(2,C) modules are locally artinian*, Proc. Amer.Math.Soc. **107** (1989), 35–37.
3. J.Dixmier, *Enveloping Algebras*, North-Holland, New York, 1977.
4. G.P.Hochschild, *Algebraic Lie algebras and representative functions*, Illinois J. Math. **3** (1959), 499–523.
5. _____, *Algebraic Lie algebras and representative functions,supplements*, Illinois J.Math. **4** (1960), 609–618.
6. J.Humphreys, *Introduction to Lie algebras and representation theory*, Springer Verlag, New York, 1972.
7. J.C.Jantzen, *Einhüllende Algebren halbeinfacher Lie-Algebren*, Springer-Verlag, Berlin, 1983.
8. T.Levasseur, *Cohomologie des algébres de Lie nilpotentes et enveloppes injectives*, Bull. Soc. Math. France **100** (1976), 377–383.
9. _____, *L'enveloppe injective du module trivial sur une algébre de Lie resoluble*, Bull.Soc.Math. France **110** (1986), 49–61.
10. _____, Private Communication 6/30/87.
11. M.E.Sweedler, *Hopf algebras*, Benjamin, New York, 1969.
12. M.Vergne, *Cohomologie des algébres de Lie nilpotentes*, Bull.Soc.Math.France **98** (1970), 81–116.

Smoothing Coherent Torsion-Free Sheaves

Amassa Fauntleroy, Department of Mathematics, North Carolina State University, Raleigh, NC 27695

Introduction

Let F be a coherent torsion free sheaf on the noetherian normal scheme X. If F has a pseudo-determinant in $Pic(X)$ (see below for definitions) then there exists a sheaf of ideals I in O_X such that if $\sigma : Z \to X$ is the blow-up of X along I, then the strict transform of F is locally free on Z.

1 Algebraic Version

Let R be a noetherian local integrally closed domain with quotient field k. Let M be a finitely generated torsion-free R-module. We shall say that M has a pseudo-determinant if there is an element $L \in Pic(R)$ and a mapping $\varphi : \wedge^d M \to L$ where $d = rk(M)$ such that φ_p is an isomorphism at all height one primes p of R. Since rank one projectives are reflexive it follows that $L \simeq (\overset{d}{\wedge} M)^{**}$. Thus M has a pseudo-determinant precisely when $[\overset{d}{\wedge} M]^{**}$ is in $Pic(R)$ and hence L is unique up to isomorphism. When M has a pseudo-determinant we define the <u>determinantal</u> <u>ideal</u> <u>of</u> <u>M</u>, I-det(M), of M as follows: If $c : L \otimes L^* \to R$ is the natural isomorphism, then I-det(M) is the image in R of the composition $\overset{d}{\wedge} M \otimes L^* \to L \otimes L^* \overset{c}{\to} R$. For any prime ideal p of height one in R, both maps in this composition become isomorphisms upon localization at p. It follows that I-det(M) is not contained in any height one prime so that ht(I-det$(M)) \geq 2$. The following lemma is well-known. We give a proof here for the convenience of the reader.

Lemma 1.1 Let Q be a prime ideal of R. Then the following are equivalent:

(i) M_q is a free R_q-module.

(ii) $\overset{d}{\wedge}(M_q)$ is a free R_q-module.

(iii) IR_q is a principal ideal where I = I-det(M).

Proof. The implications $(i) \Rightarrow (i) \Rightarrow (iii)$ being immediate we show $(iii) \Rightarrow (i)$. Fix an isomorphism λ of IR_q with R_q and let e denote the composition

$$\overset{d}{\wedge} M_q \otimes L^* \to IR_q \overset{\lambda}{\to} R_q$$

There exists $f \in L^*$ and $m_1, \ldots m_d$ in M_q such that $e(f \otimes (m_1 \wedge \ldots \wedge m_d)) = 1$. Let w_i be the $(d-1)$-form obtained from $m_1 \wedge \ldots \wedge m_d$ by deleting m_i for $1 \leq i \leq d$. Then for $m \in M_q$ we must have a relation of the form

$$bm = \sum_{i=1}^{d} b_i m_1, \quad b, b_1 \mathrm{in} R_q$$

since $m_1, \ldots, m_d$ is a basis of $M_q \otimes_{R_q} K$ over K. Then

$$
\begin{aligned}
e(f \otimes (bm \wedge w_i)) &= b \cdot e(f \otimes (m \wedge w_i)) \\
&= \sum b_j \cdot e(f \otimes m_j \wedge w_i) \\
&= \pm b_i e(f \otimes (m_1 \wedge \ldots \wedge m_d)) = \pm b_i
\end{aligned}
$$

Let $a_i = e(f \otimes m \wedge w_i)$. Then $b_i = \pm ba_i$ and $b(m - \sum \pm a_i m_i) = 0$, thus $m = \sum_i \pm a_i m_i$. Since the set $\{m_1, \ldots, m_d\}$ spans M_q and rank $M_q = d$, this set is actually a basis and M_q is free. Thus $(iii) \Rightarrow (i)$ and the proof of the lemma is complete.

Corollary. If q is a prime ideal of R at which M_q is not R_q-free, then $q \supset I$ -det(M).

Note that the above proof works under the more general setting $L \in Pic(R), \overset{d}{\wedge} M \to L$ given and the image of $\overset{d}{\wedge} M \otimes L^{-1} \to L \otimes L^{-1} \to R$ is principle. We do not actually need $L = (\overset{d}{\wedge} M)^{**}$. When R is not necessarily local we say that the R-module M has a pseudo-determinant if M_p has one over R_p for every prime ideal p of R.

2. Geometric Version

Let X be a normal variety over the algebraically closed field k and F a coherent torsion free 0_X-module of rank d. We say F has a pseudo-determinant if $\Gamma(V, F)$ has a pseudo-determinant as a $\Gamma(V, 0_X)$-module for every affine open V of X. If (L, ψ) is a pseudo-determinant we denote by $\det F$ the image under φ of $\overset{d}{\wedge} F$ and by I-$\det F$ the determinantal ideal defined locally as in Section 1. Assume now that such an F is given with pseudo-determinant (L, φ), $L \in Pic(X)$ and that F is actually reflexive and generated by its global sections.

Let $\sigma : Z \to X$ be the blow-up of X along I-$\det F$. Recall (c.f. Raynaud, Section 4.1) that for any quasi-coherent sheaf G on X, the strict transform $\hat{\sigma}(G)$ of G is the sheaf $\sigma^*(G)/N(G)$ where $N(G)$ is the subsheaf of σ^*G generated by sections whose support lies along the exceptional divisor of σ.

Lemma 2.1. If G is torsion free, then $N(G)$ is the torsion subsheaf of σ^*G and hence $\hat{\sigma}(G) = \sigma^*(G)/\text{Torsion}$.

Proof. Since $Z - V(I0_Z) = X - V(I)$ it is clear that the torsion subsheaf T of σ^*G is supported along $I0_Z$. Conversely, let $z \in Z$ be fixed and let K be the function field of Z. The exact sequence

$$0 \to N(G)_z \to (\sigma^*G)_z \to \sigma(G)_z \to 0$$

yields the sequence

$$0 \to N(G) \otimes_{0_z} K \to (\sigma^*G)_z \otimes_{0_z} K \to \hat{\sigma}(G)_z \otimes_{0_z} K \to 0$$

But the map on the right is an isomorphism since σ^*G and $\hat{\sigma}G$ agree at the generic point of Z. Thus $N(G) \otimes_{0_z} K = 0$ and $N(G) \subseteq T$.

Consider now the map $\varphi^* : \sigma^*(\overset{d}{\wedge} F) \to \sigma^*L$. If $x \in \sigma^*F$ is a torsion element and $w \in \overset{d-1}{\wedge} F$ is any $(d-1)$-form, then $\varphi^*(x \wedge w)$ is torsion hence zero. It follows that φ^* factors through $\overset{d}{\wedge} \hat{\sigma}(F)$.

Proposition 2.2. The image of $\overset{d}{\wedge} \hat{\sigma}(F)$ in σ^*L is the invertible sheaf $I0_Z \otimes \sigma^*L$.

Proof. By definition we have an exact sequence

$$\overset{d}{\wedge} F \otimes L^{-1} \to I0_X \to 0.$$

Applying the right exact functor σ^* and using the fact that φ^* factors through $\overset{d}{\wedge} \tilde{\sigma} F$ yields the exact sequence

$$\overset{d}{\wedge} \tilde{\sigma} F \otimes \sigma^* L^{-1} \to IO_Z \to 0.$$

To obtain the desired conclusion we simply tensor this last sequence with the invertible sheaf $\sigma^* L$.

Corollary 2.5. The strict transform $\hat{\sigma}(F)$ of F is locally free on Z.

Proof. The question being local, we may use the fact that $\overset{d}{\wedge} \hat{\sigma}(F)$ surjects onto a locally free sheaf and the proof of Lemma 1.1 to obtain the corollary.

Note that if X is locally factorial, every coherent torsion free sheaf has a pseudo-determinant.

2   An Example

Let $R = k[x, y, z]$ be a 3-dimensional polynomial algebra over the field k and let M be the cokernel of the map $\phi : R^2 \to R^5$ where

$$\phi = \begin{bmatrix} x & 0 \\ 0 & x \\ y & 0 \\ z & y \\ 0 & z \end{bmatrix}$$

Let m be the maximal ideal generated by x, y and z. Then since

$$0 \to R_m^2 \to R_m^5 \to M_m \to 0$$

is exact, $\text{depth}_m(M_m) = 2$ and M_m is reflexive. Since M_p is free for any prime ideal $p \neq m$ it follows that M is reflexive.

Let $\{e_i : 1 \leq i \leq 5\}$ be the standard basis for R^5. Then M is defined by the relations

$$u_1 = xe_1 + ye_3 + ze_4 = 0$$
$$u_2 = xe_2 + ye_4 + ze_5 = 0$$

Let $L \subset R^5$ be the submodule generated by u_1 and u_2 and define maps $f : \overset{2}{\wedge} L \otimes \overset{3}{\wedge} M \to \overset{5}{\wedge} R$ and $g : \overset{5}{\wedge} R \to R$ as follows:

$$g(e_1 \wedge \ldots \wedge e_5) = 1$$

$$f(u_1 \wedge u_2 \otimes m_1 \wedge m_2 \wedge m_3) = u_1 \wedge u_2 \wedge v_1 \wedge v_2 \wedge v_3$$

where v_i is any preimage of m_i in R^5. If m_1 has preimages v and v' then $v - v' = au_1 + bu_2$ with a, b in R. Hence

$$u_1 \quad \wedge \quad u_2 \wedge v \wedge v_2 \wedge v_3 =$$

$$= \quad u_1 \wedge u_2 \wedge (v' + au_1 + bu_2) \wedge v_2 \wedge v_3$$

$$= \quad u_1 \wedge u_2 \wedge v' \wedge v_2 \wedge v_3$$

It follows from similar calculations using m_2, m_3 that f is well defined. Let $\psi = g \circ f$. Then ψ induces an isomorphism

$$\overset{2}{\wedge} L_p \otimes (\overset{3}{\wedge} M)_p \to R_p$$

for all height one primes p of R. Thus $\overset{2}{\wedge} L \simeq R$ is identified with the dual of $\left[\overset{3}{\wedge} M \right]^{**}$. Computing $\varphi(u_1 \wedge u_2 \otimes e_i \wedge e_j \wedge e_k)$ for $i < j < k$ we find I-det$(M) = I = (x, y, z)^2$. This is just the ideal generated by the 2×2 minors of ϕ.

Let $\sigma : Z \to X = \operatorname{Spec} R$ be the blow-up of X along the sheaf of ideals IO_X. Since $I = m^2$, Z is isomorphic to the blow-up of X at the point $(0, 0, 0)$. The scheme Z is covered by three open affines Z_x, Z_y and Z_x where, for example,

$$\Gamma(Z_x, 0_Z) = R[\alpha, \beta], x\alpha = y, x\beta = z.$$

A similar description can be given for the other two open affines Z_y, Z_z. Note that M_x is generated by elements of the form ℓ/x where ℓ is a linear form in $e_1, \ldots, e_5$. On the affine open set z_x, $\hat{\sigma}(M) = E_0$ is generated by e_3, e_4 and e_5. Similarly, $\hat{\sigma}(M)$ is free of rank 3 over Z_y and Z_z.

References

[1] M. Raynaud, *Flat modules in algebraic geometry*, Compositio Math. 24 (1972), pp. 11–31.

Projective Covers and Quasi-Isomorphisms

MARK A. GODDARD Department of Mathematical Sciences, The University of Akron, Akron, Ohio 44325-4002

1. Introduction

In this paper, the concept of the projective cover of a module is generalized to the case of complexes of finitely generated modules over a Noetherian local ring. The projective cover of a complex possesses many but not all of the key properties of the projective cover in the module case. Using the fact that the projective cover is a quasi-isomorphism, we see that the projective cover is closely related to the free resolution and minimal free resolution of a complex (see Roberts, 1980).

We begin with some introductary terminology. In this paper, R will be a commutative ring with identity. A complex C of R-modules is a sequence of R-module homomorphisms

$$\ldots \to C_n \xrightarrow{\delta_n} C_{n-1} \to \ldots \to C_1 \xrightarrow{\delta_1} C_0 \to \ldots$$

satisfying $\delta_{i-1} \circ \delta_i = 0$ for all $i \in Z$. The maps δ_i are called the **boundary maps** of the complex C. If $\ker \delta_{i-1} = \operatorname{im} \delta_i$ for all $i \in Z$ then C is called an **exact sequence**. If $C_i = 0$ for i sufficiently small then C is said to be **bounded above**. We will assume that if C is bounded above then $C_i = 0$ for $i < 0$.

A **map of complexes** $C \xrightarrow{\phi} D$ is a sequence of R-module homomorphisms $C_i \xrightarrow{\phi_i} D_i$ which commutes with the boundary maps of the complexes C and D. A map of complexes ϕ will be said to be **surjective** if each map ϕ_i is surjective. A **quasi-isomorphism** is a map of complexes $C \xrightarrow{\phi} D$ which induces an isomorphism on homology, i.e. each induced map $H_i(C) \xrightarrow{\phi_i^*} H_i(D)$ is an isomorphism.

Generalizing the definition of a projective R-module, we define a projective complex to be a complex **Q** satisfying the condition that for any complexes **C** and **D**, any map of complexes **Q** $\xrightarrow{\phi}$ **C** and any surjective map of complexes **D** $\xrightarrow{\epsilon}$ **C** the diagram

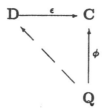

can be completed by a map of complexes.

It is worth noting that every complex of projective R-modules does not form a projective complex. (See Goddard (1993) for an example.) Nonetheless, the collection of all projective complexes can be easily described as seen in the following proposition:

PROPOSITION 1.1 *A bounded above complex* **P** *is a projective complex if and only if* **P** *is an exact sequence of projective modules.*

Proof: See Roberts (1986), proposition 2.1. □

By a generalization of the categorical definition of the projective cover of a module, we now define the projective cover of a complex:

DEFINITION: The projective cover of a complex **C** is a complex **P** of projective modules and a map of complexes **P** $\xrightarrow{\epsilon}$ **C** satisfying the following two conditions:

1. For any complex **Q** of projective modules, and any map of complexes **Q** $\xrightarrow{\phi}$ **C**, the diagram

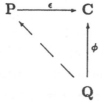

   can be completed by a map of complexes.

2. The diagram

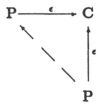

can only be completed by maps of complexes which are automorphisms of P_i in each degree.

Note that a complex of projective modules satisfying only condition 1 of the definition above is called a projective precover of **C**.

2. Properties of the Projective Cover

In this section, we shall consider some of the important properties of the projective cover of a module and compare and contrast these with the properties of the projective cover of a complex. The projective cover $P \xrightarrow{\epsilon} M$ of a module, M, is a unique surjective map from a projective module P onto M. The existence of the projective cover may be shown in the case where M is a finitely generated module over a Noetherian local ring. We will see that the existence and uniqueness of projective covers is preserved in the case of complexes as is the surjective property. The projective cover of a complex need not be a projective complex however. For an example of a projective cover which is not itself projective, see Goddard (1993).

We begin with the proof of the existence of the projective cover of a complex. Although the theorem below is stated for finitely generated modules over a Noetherian local ring, we may replace this with modules over a perfect ring or any other condition guaranteeing the existence of projective covers in the module case.

THEOREM 2.1 *If* **C** *is a bounded above complex of finitely generated modules over a Noetherian local ring then* **C** *has a projective cover.*

Proof: Let $P_0 \xrightarrow{\epsilon_0} C_0$ be the projective cover of C_0 whose existence is guaranteed by the fact that C_0 is a finitely generated module over a Noetherian local ring and let $P_i = 0$ for all $i < 0$. From this point we proceed inductively to construct the projective cover of the remainder of the complex. Assuming that we have completed the ith step

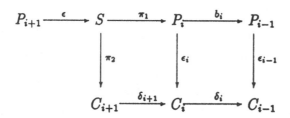

we make use of the standard pullback construction given by the set

$$S = \{(p,c) \in ker(b_i) \oplus C_{i+1} | \delta_{i+1}(c) = \epsilon_i(p)\}.$$

Since S is a submodule of $P_i \oplus C_{i+1}$ and hence is itself finitely generated, we can construct the projective cover $P_{i+1} \xrightarrow{\epsilon} S$ of S to obtain the commutative diagram

$$
\begin{array}{ccccccc}
P_{i+1} & \xrightarrow{\epsilon} & S & \xrightarrow{\pi_1} & P_i & \xrightarrow{b_i} & P_{i-1} \\
& & \downarrow{\pi_2} & & \downarrow{\epsilon_i} & & \downarrow{\epsilon_{i-1}} \\
& & C_{i+1} & \xrightarrow{\delta_{i+1}} & C_i & \xrightarrow{\delta_i} & C_{i-1}
\end{array}
$$

If we define $b_{i+1} = \pi_1 \circ \epsilon$ and $\epsilon_{i+1} = \pi_2 \circ \epsilon$ then this construction clearly yields a complex $\mathbf{P}$ of projective modules and a map of complexes $\mathbf{P} \xrightarrow{\epsilon} \mathbf{C}$. All that remains to be shown is that this map satisfies properties 1 and 2 stated in the definition of the projective cover.

First we let $\mathbf{Q} \xrightarrow{\phi} \mathbf{C}$ be a map of complexes from a complex of projective modules $\mathbf{Q}$ into $\mathbf{C}$. To satisfy condition 1, we must find a map of complexes $\mathbf{Q} \xrightarrow{h} \mathbf{P}$ so that $\epsilon \circ h = \phi$. For $i < 0$ we let h_i equal the zero map and we define h_0 to be the map completing the diagram

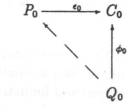

From this point we proceed inductively to construct the map of complexes. Let us assume that we have already defined h_j for $j \le i$ so that $\mathbf{h}$ commutes with the appropriate boundary maps and satisfies $\epsilon_j \circ h_j = \phi_j$ for all $j \le i$. To construct h_{i+1} we consider the diagram

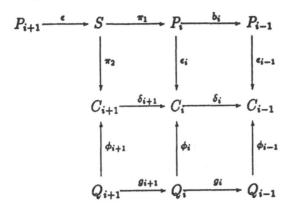

Define $f : Q_{i+1} \to S$ by $f(q) = (h_i \circ g_{i+1}(q), \phi_{i+1}(q))$. Since P_{i+1} is the projective cover of S, there exists a map $Q_{i+1} \overset{h_{i+1}}{\to} P_{i+1}$ such that $\epsilon \circ h_{i+1} = f$. It follows easily that $\epsilon_{i+1} \circ h_{i+1} = \phi_{i+1}$ and that the map h defined in this manner still commutes with the boundary maps.

Next, let $P \overset{h}{\to} P$ be a map of complexes satisfying $\epsilon \circ h = \epsilon$. We must show that each map $P_i \overset{h_i}{\to} P_i$ is an automorphism of P_i. For $i < 0$ this is clear since each map h_i is the zero map and since $P_0 \overset{\epsilon_0}{\to} C_0$ is the projective cover of C_0, we know that h_0 is an automorphism as well. Proceeding inductively, let us assume that h_j is an automorphism for all $j \leq i$ and show that h_{i+1} must be an automorphism as well.

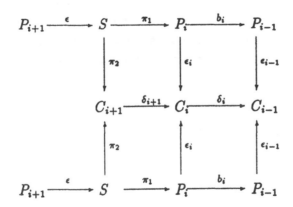

We can define an automorphism $S \overset{f}{\to} S$ by $f(p, c) = (h_i(p), c)$. It can easily be seen that f makes the diagram

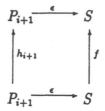

commute and since $P_{i+1} \xrightarrow{\epsilon} S$ is the projective cover of S, it follows easily that h_{i+1} is an automorphism of P_{i+1} and our proof is complete. $\square$

In the same way that the definition of a projective cover can be generalized to yield an injective or flat cover in the case of modules (Enochs, 1981), we can also define the injective or flat cover of a complex. Proceeding exactly as in the proof of the last theorem, we can prove the existence of the injective cover of any bounded above complex of modules over a Noetherian ring.

Next we consider the question of uniqueness. By generalizing the proof used in the module case, we can easily prove that the projective cover of a complex is unique up to isomorphism. The result below holds for injective covers of complexes as well.

PROPOSITION 2.1 *The projective cover of a complex* **C** *is unique up to isomorphism.*

Proof: Let $\mathbf{P} \xrightarrow{\epsilon} \mathbf{C}$ and $\mathbf{Q} \xrightarrow{\phi} \mathbf{C}$ be projective covers of **C**. Since **P** is a projective cover, there exists a map of complexes $\mathbf{Q} \xrightarrow{h} \mathbf{P}$ such that $\epsilon \circ h = \phi$. Similarly since **Q** is a projective cover, there exists a map $\mathbf{P} \xrightarrow{k} \mathbf{Q}$ such that $\phi \circ k = \epsilon$. The map $h_i \circ k_i$ must be an automorphism of P_i and $k_i \circ h_i$ must be an automorphism of Q_i for each i. Thus the map $P_i \xrightarrow{k_i} Q_i$ is an automorphism for each i. $\square$

Two other important properties of the projective cover of a module are the facts that it is a surjective map and that it is preserved by direct sums. As we shall see below, these properties are retained by projective covers of complexes. We begin with a useful lemma:

LEMMA 2.1 *If* **C** *is a bounded above complex of finitely generated modules over a Noetherian local ring then there exists a map of complexes* $\mathbf{F} \xrightarrow{\phi} \mathbf{C}$ *such that*

1. *Each map* $F_i \xrightarrow{\phi_i} C_i$ *is surjective.*

2. F_i *is a free module for each* i.

3. ϕ *is a quasi-isomorphism.*

4. $\phi_i(Z_i(\mathbf{F})) = Z_i(\mathbf{C})$ *for each* i.

Proof: Roberts (1980), pp 42-44. $\square$

THEOREM 2.2 *If* $P \xrightarrow{\epsilon} C$ *is a projective cover of a bounded above complex of finitely generated modules over a Noetherian local ring then* ϵ_i *is a surjection for all* i.

Proof: By lemma 2.1, there exists a map of complexes $F \xrightarrow{\phi} C$ where each module F_i is free and each map ϕ_i is surjective. Since P is the projective cover of C and F is a complex of projective modules, the diagram

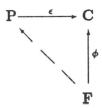

can be completed by a map of complexes $F \xrightarrow{h} P$. Since each ϕ_i is surjective, each map ϵ_i must be surjective as well. □

PROPOSITION 2.2 *If* $P^i \xrightarrow{\epsilon^i} C^i$ *are projective covers of complexes for* $1 \leq i \leq n$ *then* $\oplus P^i \xrightarrow{\oplus \epsilon^i} \oplus C^i$ *is the projective cover of* $\oplus C^i$.

Proof: We will show the result for $n = 2$. The general result follows by induction. Let $P' \xrightarrow{\epsilon'} C'$ and $P'' \xrightarrow{\epsilon''} C''$ be projective covers of C' and C'' respectively. We need to show that the map of complexes $P' \oplus P'' \xrightarrow{\epsilon' \oplus \epsilon''} C' \oplus C''$ satisfies properties 1 and 2 of the projective cover.

Given a map of complexes $Q \xrightarrow{\phi} C' \oplus C''$ with each Q_i projective, in order to satisfy property 1, we need a map of complexes $Q \xrightarrow{h} P' \oplus P''$ such that $(\epsilon' \oplus \epsilon'') \circ h = \phi$. Since P' and P'' are projective covers, there are maps h' and h'' such that $\epsilon' \circ h' = \pi_1 \circ \phi$ and $\epsilon'' \circ h'' = \pi_2 \circ \phi$. Let $h = h' \oplus h''$ and we have satisfied condition 1.

Next we need to argue that any map $P' \oplus P'' \xrightarrow{h} P' \oplus P''$ satisfying $(\epsilon' \oplus \epsilon'') \circ h = (\epsilon' \oplus \epsilon'')$ must be an automorphism in each degree. If we write the map h in matrix form as

$$h = \begin{pmatrix} a & c \\ b & d \end{pmatrix}$$

then since P' and P'' are projective covers, the maps a and d must be automorphisms of P' and P'' respectively. Furthermore, the map $k = a - c \circ d^{-1} \circ b$ completes the diagram

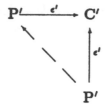

and hence must be an automorphism of $\mathbf{P}'$. Thus the map

$$j = \begin{pmatrix} k & c \circ d^{-1} \\ 0 & id \end{pmatrix}$$

is an automorphism as well. Since

$$m = \begin{pmatrix} a^{-1} & 0 \\ 0 & d^{-1} \end{pmatrix}$$

and

$$n = \begin{pmatrix} a & 0 \\ -b & id \end{pmatrix}$$

are automorphisms and $j = h \circ m \circ n$, it follows that h must also be an automorphism. $\square$

Under certain conditions, this result may be applied not only to finite direct sums but to countably infinite direct sums as well.

PROPOSITION 2.3 *Let* $k^* \in Z$ *be fixed. Let* $\{C^i\}_{i \in Z^+}$ *be a collection of bounded above complexes of modules such that* $C_j^i = 0$ *for all* $j < k^*$. *If* $\mathbf{P}^i \overset{\epsilon^i}{\to} \mathbf{C}^i$ *are projective covers for all* $i \in Z^+$, *then* $\oplus \mathbf{P}^i \overset{\oplus \epsilon^i}{\to} \oplus \mathbf{C}^i$ *is a projective cover if and only if for each sequence* $1 \leq k_1 < k_2 < \dots$ *of positive integers, corresponding maps* $\mathbf{P}^{k_n} \overset{f^n}{\to} \mathbf{P}^{k_{n+1}}$ *satisfying* $f_n(\mathbf{P}^{k_n}) \subseteq ker(\epsilon^{k_{n+1}})$ *and for each* $x \in P_k^{k_1}$ *there exists an* $m \geq 1$ *such that* $f_k^m \circ f_k^{m-1} \circ \dots \circ f_k^1(x) = 0$.

Proof: The proof follows immediately from the similar result for projective covers of modules proved by Enochs (1981). $\square$

3. Projective Covers and Related Quasi-Isomorphisms

In this section, we will see that the projective cover has an additional property which is not inherited from the module case: The projective cover $\mathbf{P}$ of a complex $\mathbf{C}$ retains the homology of the original complex. Before we prove this result, we first consider two other important examples of quasi-isomorphisms defined by Roberts (1980). First of all, a free resolution of a complex $\mathbf{C}$ is simply a quasi-isomorphism from a complex $\mathbf{F}$ of free modules into $\mathbf{C}$. We shall refer to a free resolution which is also a surjective map of complexes as a surjective free resolution.

While the surjective free resolution of a bounded above complex of finitely generated modules over a Noetherian local ring is guaranteed to exist by lemma 2.1, it is not generally unique. In order to obtain uniqueness, Roberts defines a new class of free resolutions which are no longer necessarily surjective.

DEFINITION: A minimal free resolution of a complex C is a free resolution $F \xrightarrow{\phi} C$ with the property that each map $F_i \xrightarrow{\phi_i} C_i$ is defined by a matrix with entries in the maximal ideal m of the local ring R.

Under exactly the same conditions which guarantee the existence of the projective cover and surjective free resolution, the existence and uniqueness of the minimal free resolution have been shown. Furthermore, Roberts has proven that every free resolution is the direct sum of the minimal free resolution and an exact complex. In order to clarify the relationship between the projective cover and these two new types of quasi-isomorphisms, we begin with the following lemma.

LEMMA 3.1 *Let C be a bounded above complex. If $P \xrightarrow{\epsilon} C$ is a surjective quasi-isomorphism then P is a projective precover of C*

Proof: Let Q be a complex of projective modules and let $Q \xrightarrow{\phi} C$ be a map of complexes. We need to find a map $Q \xrightarrow{h} P$ so that $\epsilon \circ h = \phi$. We let $h_i = 0$ for $i < 0$. For the $i = 0$ case, we first need to note that for all $i \geq 0$, $\epsilon_i(Z_i(P)) = Z_i(C) = \{c \in C_i | \delta_i(c) = 0\}$. In order to do so, let $c \in Z_i(C)$. Since ϵ is a quasi-isomorphism, there is a $p \in Z_i(P)$ such that $\epsilon_i(p) - c \in B_i(C)$. Since ϵ is surjective and commutes with the boundary maps, there exists $p' \in P_{i+1}$ such that $\epsilon_i(b_{i+1}(p')) = \delta_{i+1}(\epsilon_{i+1}(p')) = \epsilon_i(p) - c$. Thus $c = \epsilon_i(p - b_{i+1}(p')) \in \epsilon_i(Z_i(P))$.

Since $\epsilon_0(Z_0(P)) = Z_0(C)$ and Q_0 is projective, the diagram

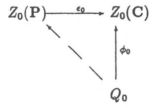

can be completed by a map h_0. We now proceed inductively to construct h. Assume that we have constructed h_j such that $\epsilon_j \circ h_j = \phi_j$ for all $j \leq i$ and h commutes with the boundary maps. We now wish to construct h_{i+1}.

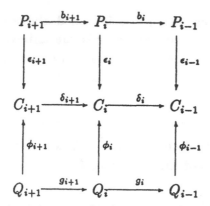

Since $\epsilon_i \circ h_i \circ g_{i+1}(q) \in B_i(\mathbf{C})$ for all $q \in Q_{i+1}$ and ϵ is a quasi-isomorphism, we have $h_i \circ g_{i+1}(q) \in B_i(\mathbf{P})$ for all $q \in Q_{i+1}$. Thus the diagram

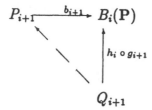

can be completed by a map h' since Q_{i+1} is projective. For $q \in Q_{i+1}$, we have $(\epsilon_{i+1} \circ h' - \phi_{i+1})(q) \in Z_{i+1}(\mathbf{C})$ so the diagram

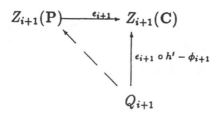

can be completed by a map h'' since the module Q_{i+1} is projective and the map $Z_{i+1}(\mathbf{P}) \overset{\epsilon_{i+1}}{\to} Z_{i+1}(\mathbf{C})$ is surjective. If we let $h_{i+1} = h' - h''$ then it follows easily that h commutes with the boundary and $\epsilon_{i+1} \circ h_{i+1} = \phi_{i+1}$. $\square$

THEOREM 3.1 *Let* $\mathbf{C}$ *be a bounded above complex of finitely generated modules over a Noetherian local ring. If* $\mathbf{P} \overset{\epsilon}{\to} \mathbf{C}$ *is the projective cover of* $\mathbf{C}$ *then* ϵ *is a quasi-isomorphism.*

Proof: By lemma 2.1, we know that there exists a surjective map of complexes $\mathbf{F} \overset{\phi}{\to} \mathbf{C}$ where each module F_i is free, $\phi_i(Z_i(\mathbf{F})) = Z_i(\mathbf{C})$ for all i and ϕ is a quasi-isomorphism.

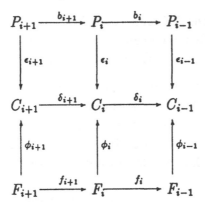

Since $\mathbf{P}$ is the projective cover of $\mathbf{C}$, there exists a map of complexes $\mathbf{F} \xrightarrow{h} \mathbf{P}$ such that $\epsilon \circ h = \phi$. Using this map h, we will prove that $H_i(\mathbf{P}) = H_i(\mathbf{C})$.

First let $c \in \ker(\delta_i)$. There exists $q \in \ker(f_i)$ such that $\phi_i(q) = c$. It can easily be seen that $h_i(q)$ lies in $\ker(b_i)$ and that $\epsilon_i(h_i(q)) = c$. Thus $\epsilon_i(Z_i(\mathbf{P})) = Z_i(\mathbf{C})$.

Since the diagram commutes, $\epsilon_i(B_i(\mathbf{P})) \subseteq B_i(\mathbf{C})$. Thus all that remains to be shown is that $B_i(\mathbf{C}) \subseteq \epsilon_i(B_i(\mathbf{P}))$. Let $p \in \ker(b_i)$ such that $\epsilon_i(p) = \delta_{i+1}(c)$ for some $c \in C_{i+1}$. All we need to show is that $p = b_{i+1}(p')$ for some $p' \in P_{i+1}$.

Since ϕ is a surjective quasi-isomorphism, by the previous lemma, $\mathbf{F} \xrightarrow{\phi} \mathbf{C}$ is a projective precover of $\mathbf{C}$. Thus the diagram

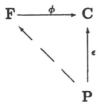

can be completed by a map of complexes, k. Since $\epsilon \circ h \circ k = \epsilon$ and $\mathbf{P}$ is a projective cover, it follows that $h \circ k$ is an automorphism of $\mathbf{P}$. Let $q = k_i((h_i \circ k_i)^{-1}(p))$. With this choice, we have $q \in \ker(f_i)$ and $\phi_i(q) \in \delta_{i+1}(C_{i+1})$. Since ϕ is a quasi-isomorphism, we know that there exists $q' \in F_{i+1}$ such that $f_{i+1}(q') = q$. Since h is a map of complexes, $b_{i+1} \circ h_{i+1}(q') = h_i(q) = p$ and we have shown that $p \in B_i(\mathbf{P})$. $\square$

Since the projective covers of modules used in the proof of theorem 2.1 are, in fact, all free modules, we can see by theorem 2.2 and theorem 3.1 that the projective cover of a complex is in fact an example of a surjective free resolution (though it is not the same surjective free resolution which is shown by Roberts to exist in lemma 2.1). By invoking proposition 1.1, we thus have the following corollary:

COROLLARY 3.1 *If* C *is a bounded above complex of finitely generated modules over a Noetherian local ring then the projective cover of* C *is equal to the direct sum of the minimal free resolution of* C *and a projective complex.*

In many examples, the projective cover and the minimal free resolution of a complex turn out to be identical. (See Goddard (1993) for an example where the projective cover differs from the minimal free resolution.) There are a number of conditions which guarantee that the projective cover and minimal free resolution are identical. For example, if the complex C has the form

$$C = \ldots \to 0 \to C_n \to C_{n-1} \to \ldots \to C_1 \to C_0 \to 0 \to \ldots$$

where $C_0, C_1, \ldots, C_{n-2}$ are all free modules then the projective cover of C is equal to the minimal free resolution of C if and only if $\delta_i(C_i) \subseteq m C_{i-1}$ for all i. We conclude this paper with a theorem which specifies the most straightforward condition which can be placed upon the minimal free resolution to guarantee that it is identical to the projective cover. We first need the following pair of lemmas.

LEMMA 3.2 *If* F $\to$ C *and* G $\to$ C *are two minimal free resolutions of* C *and* G $\xrightarrow{h}$ F *is a map of complexes making the diagram*

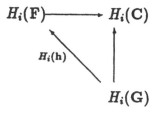

commute for all values of i, *then each map* h_i *is an isomorphism.*

Proof: See Roberts (1980). □

LEMMA 3.3 *If* F $\xrightarrow{\phi}$ C *is the minimal free resolution of* C *then the diagram*

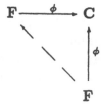

can only be completed by maps of complexes which are automorphisms of F_i *in each degree.*

Proof: If h is a map of complexes completing the diagram above, then $H_i(\phi) \circ H_i(h) = H_i(\phi)$ as well and by lemma 3.2, h is an isomorphism from **F** onto itself. $\square$

THEOREM 3.2 *The minimal free resolution of a complex* **C** *is identical to the projective cover of* **C** *if and only if the minimal free resolution is a surjective map of complexes.*

Proof: Since the projective cover is a surjective map of complexes, the first implication is obvious. To proceed in the other direction, let us assume that the minimal free resolution $\mathbf{F} \xrightarrow{\phi} \mathbf{C}$ is surjective. By lemma 3.3, the minimal free resolution always satisfies the second defining condition of the projective cover. Thus, all we need to show is that the minimal free resolution is a projective precover. This follows immediately from lemma 3.1 since the minimal free resolution is a quasi-isomorphism and is surjective by assumption. $\square$

REFERENCES

1. Enochs, E. (1981). *Injective and Flat Covers, Envelopes, and Resolvents*, Israel Journal of Mathematics, (3)39, pp. 189-209.

2. Goddard, M. (1993). *Projective Covers of Complexes*, Proceedings of the 1992 Ohio State - Denison Mathematics Conference.

3. Roberts, P. (1980). *Homological Invarients of Modules over Commutative Rings*, Les Presses de l'Université de Montreal.

4. Roberts, P. (1986) *Some Remarks on the Homological Algebra of Multiple Complexes*, Journal of Pure and Applied Algebra, 43, pp. 99-110.

5. Rotman, J. (1979) *An Introduction to Homological Algebra*, Academic Press Inc., New York.

On Dihedral Algebras and Conjugate Splittings

DARRELL E. HAILE   Department of Mathematics, Indiana University, Bloomington, Indiana, haile@ucs.indiana.edu

Let F be a field and let D be an F–central division algebra. A well known theorem of Wedderburn (1921) says that if an irreducible polynomial $f(x) \in F[x]$ has a root θ in D then $f(x)$ decomposes into a product of linear factors in $D[x]$. More precisely there are elements $\theta_1 = \theta, \theta_2, \ldots, \theta_k$ in D, such that each θ_i is conjugate to θ and $f(x) = (x - \theta_k)(x - \theta_{k-1}) \cdots (x - \theta_2)(x - \theta_1)$. Note: The elements θ_i need not commute! In the same paper Wedderburn proved that if the dimension of D over F is nine, then one can do better. In that case if θ is in $D - F$ then $f(x)$ (which will be of degree three) decomposes over D in a very special way: there is an element y in D such that $y^3 \in F^\times$ and $f(x) = (x - y^{-2}\theta y^2)(x - y^{-1}\theta y)(x - \theta)$. Wedderburn used this fact to prove that every division algebra of dimension nine is cyclic, that is has a maximal subfield (necessarily of degree three over F) that is a cyclic extension of F.

From this point on the field F will be assumed infinite. In Haile (1989, 1991) I investigated the possibility of extending this result on polynomials to division algebras of arbitrary dimension. This inquiry was spurred in part by the discovery that such a decomposition is related to other questions of interest and in particular to the theory of Clifford algebras, which will now be described. If $g(u_1, u_2, \ldots, u_n) \in F[u_1, \ldots, u_n]$ is a homogeneous polynomial of degree m (say) then the *Clifford algebra* C_g of g over F is defined to be the algebra $F\{x_1, \ldots, x_n\}/I$, where $F\{x_1, \ldots, x_n\}$ is the free algebra on n variables and I is the ideal generated by the set $\{(a_1 x_1 + a_2 x_2 + \cdots + a_n x_n)^m - g(a_1, \ldots, a_n) | a_1, \ldots, a_n \in F\}$. Of course if g has degree two, then this is the classical Clifford algebra. For more information in the general case the reader is referred to the bibliography, particularly [Childs (1978), Tesser (1988), Haile & Tesser (1988), Hodges & Tesser (1989)] and their references. The connection with decompositions of the type described above is as follows: Suppose A is an F–algebra and $f(x) \in F[x]$. We will say $f(x)$ has a *conjugate splitting* in A if there is an invertible element y in A and an element θ in A such that $y^n \in F$ and $f(x) = (x - y^{-(n-1)}\theta y^{n-1})(x - y^{-(n-2)}\theta y^{n-2}) \cdots (x - y^{-1}\theta y)(x - \theta)$ in $A[x]$. We call the scalar y^n the *associated constant* for the splitting. Let $g(u_1, u_2) = u_2^n f(u_1/u_2)$ be the homogenization of f. Then one has the following theorem Haile (1991).

The following conditions are equivalent.

1. The polynomial $f(x)$ has a conjugate splitting in A with associated constant α.
2. There is an F-algebra homomorphism from $C_{\alpha g}$ to A.

If one restricts the algebras A in which the conjugate splitting is to occur, then there is an interesting connection with function fields of curves. Namely let $f(x)$ and $g(u_1, u_2)$ be as above and let n denote the degree of f. Let Y be the curve given by the equation $z^n - \alpha g(u_1, u_2)$ and let $F(Y)$ denote its function field. If the algebra A is central simple over F and of degree n over F, then conditions (1) and (2) above are equivalent to the following condition:

3. The algebra A is split by $F(Y)$ (that is, $A \otimes_F F(Y) \cong M_n(F(Y))$).

As an example of the usefulness of this result, consider the case where f is an irreducible cubic polynomial. In that case the ring $C_{\alpha g}$ is well understood (See Haile (1984)). It is an Azumaya algebra of rank nine over its center Z and Z is the affine ring of the elliptic curve $E : S^2 = R^3 - 27\Delta$ where $\Delta \in F$ is the discriminant of the form αg. It follows that the finite dimensional images of $C_{\alpha g}$ are simple algebras of degree 3 over their centers and these images are in one-to-one correspondence with the points of the curve E. So the conjugate splittings are in this sense parametrized by the points on this curve. For explicit examples the reader is referred to Haile (1991).

The purpose of this paper is to present another situation in which conjugate splittings can be proved to arise. Let n be an odd integer and let F be a field containing a primitive $n-th$ root of unity, ζ. Assume F admits a Galois extension L/F such that $Gal(L/F) \cong D_n$, the dihedral group of order $2n$. Let $\sigma, \tau \in Gal(L/F)$ such that $\sigma^n = \tau^2 = \tau\sigma\tau\sigma = 1$. Let K be the fixed field of τ and let L_0 be the fixed field of σ. We are interested in central simple F-algebras of degree n that are split by L. These are the *dihedral algebras* of the title. Because n is odd, any such algebra is in fact split by K, and so contains a maximal subfield F-isomorphic to K. Dihedral algebras have been studied by Rowen and Saltman (1982), who proved that such algebras always have cyclic maximal subfields. This fact was reproved using different techniques by Mammone and Tignol (1987). The strategy of the Rowen-Saltman paper is to prove that if A is dihedral then there is an element $a \in A$ such that $a^n \in F$ but no smaller positive power of a is in F. Because F contains a primitive $n - th$ of unity, the extension $F[a]$ is the desired cyclic maximal subfield.

We need more notation. The extension L/L_0 is cyclic of degree n and so there is an element $\alpha \in L$ such that $L = L_0(\alpha)$ and $\sigma(\alpha) = \zeta\alpha$. We may assume $\tau(\alpha) = \alpha^{-1}$ (If not then $\sigma(\alpha/\tau(\alpha)) = \zeta^2(\alpha/\tau(\alpha))$ and so we may replace α by $(\alpha/\tau(\alpha))^{n+1}$). Let $\alpha^n = a \in L_0$. Then $\alpha + \alpha^{-1} \in L_0$, so $a + a^{-1} \in F$. The algebra $A \otimes_F L_0$ contains L as a cyclic maximal subfield. By the Skolem-Noether theorem there is an element $\beta \in A \otimes_F L_0$ such that $\alpha\beta = \zeta\beta\alpha$ and $\beta^n \in L_0$. We may extend τ to $A \otimes_F L_0$ by letting it act on the second factor. It is shown in Lemma 2 of Rowen and Saltman (1982) that we may assume $\tau(\beta) = \beta^{-1}$. Hence $\beta + \beta^{-1} \in A$. Our main tool is a strengthening of Lemma 3 of Rowen and Saltman (1982). We need the following result on matrices. The statement is probably well known, but for convenience we include a proof.

LEMMA 1. Let F be a field and let A be the following $n \times n$ matrix over F:

$$
\begin{pmatrix}
0 & a_1 & 0 & 0 & 0 & \cdots & 0 & b_{n-1} \\
b_0 & 0 & a_2 & 0 & 0 & \cdots & 0 & 0 \\
0 & b_1 & 0 & a_3 & 0 & \cdots & 0 & 0 \\
0 & 0 & b_2 & 0 & a_4 & \cdots & 0 & 0 \\
\vdots & \vdots & \vdots & \vdots & \vdots & \cdots & \vdots & \vdots \\
0 & 0 & 0 & 0 & 0 & \cdots & 0 & a_{n-1} \\
a_0 & 0 & 0 & 0 & 0 & \cdots & b_{n-2} & 0
\end{pmatrix}
$$

If $f(x) = \sum_{k=0}^{n} c_k x^k$ is the characteristic polynomial of A, then $c_{n-(2k+1)} = 0$ for all k such that $0 < n - (2k+1) < n$.

Proof. We proceed by induction on n. Let $h_n(a_0, a_1, \ldots, a_{n-1}; b_0, b_1, \ldots, b_{n-1}) \in F$ denote the determinant of $xI - A$ (that is, this is just another notation for $f(x)$). Expanding along the first row we obtain $h_n(a_0, a_1, \ldots, a_{n-1}; b_0, b_1, \ldots, b_{n-1}) = x h_{n-1}(0, a_2, a_3, \ldots, a_{n-1}; b_1, b_2, \ldots, b_{n-2}, 0) + a_1 S + b_{n-1} T$ for appropriate cofactors S and T. Expanding these cofactors along their first columns gives $h_n(a_0, a_1, \ldots, a_{n-1}; b_0, b_1, \ldots, b_{n-1}) = x h_{n-1}(0, a_2, a_3, \ldots, a_{n-1}; b_1, b_2, \ldots, b_{n-2}, 0) - a_1 b_0 h_{n-2}(0, a_3, a_4, \ldots, a_{n-1}; b_2, b_3 \ldots, b_{n-2}, 0) - a_0 b_{n-1} h_{n-2}(0, a_2, a_3, \ldots, a_{n-2}; b_1, b_2, \ldots, b_{n-3}, 0) + (-1)^n a_0 a_1 \cdots a_{n-1} + (-1)^n b_0 b_1 \cdots b_n - 1$. The result follows immediately by induction. $\square$

PROPOSITION 2. Let $c, d, e, f \in L_0$. If $c\alpha + d\alpha^{-1}$ and $e\beta + f\beta^{-1}$ are invertible, then $\eta^n \in L_0$, where $\eta = (c\alpha + d\alpha^{-1})(e\beta + f\beta^{-1})^{-1}$.

Proof. We pass to $\overline{A} = A \otimes_F \overline{F}$, where $\overline{F}$ denotes the algebraic closure of F, which we may assume contains L_0. Let $r \in \overline{F}$ such that $r^n = b (= \beta^n)$. Then $\beta = \overline{\beta} r$ where $\overline{\beta}^n = 1$. The algebra $\overline{A}$ is generated as an $\overline{F}$-algebra by α and $\overline{\beta}$. Moreover $\overline{A}$ acts on the n-dimensional vector space $\overline{F}[\alpha]$ via $\sum_{i,j} s_{ij} \alpha^i \overline{\beta}^j \cdot \alpha^k = \sum_{i,j} s_{ij} \zeta^{jk} \alpha^{i+k}$. In particular if we use the powers of α as a basis of $\overline{F}[\alpha]$, the element $e\beta + f\beta^{-1}$ acts diagonally, in fact $(e\beta + f\beta^{-1}) \cdot \alpha^k = (\zeta^k er + \zeta^{-k} fr^{-1})\alpha^k$. It follows that the matrix of η is as follows:

$$
\begin{pmatrix}
0 & a_1 & 0 & 0 & 0 & \cdots & 0 & ab_{n-1} \\
b_0 & 0 & a_2 & 0 & 0 & \cdots & 0 & 0 \\
0 & b_1 & 0 & a_3 & 0 & \cdots & 0 & 0 \\
0 & 0 & b_2 & 0 & a_4 & \cdots & 0 & 0 \\
\vdots & \vdots & \vdots & \vdots & \vdots & \cdots & \vdots & \vdots \\
0 & 0 & 0 & 0 & 0 & \cdots & 0 & a_{n-1} \\
a^{-1}a_0 & 0 & 0 & 0 & 0 & \cdots & b_{n-2} & 0
\end{pmatrix}
$$

where $a_i = c(\zeta^i er + \zeta^{-i} fr^{-1})^{-1}$ and $b_i = d(\zeta^i er + \zeta^{-i} fr^{-1})^{-1}$. By Lemma 1 the characteristic polynomial $f(x) = \sum_{k=0}^{n} c_k x^k$ (say) of η has the property that $c_k = 0$ for all even positive k. If we reverse the roles of α and β then the argument just given shows that, except for the constant term, the even coefficients of the characteristic polynomial of η^{-1} are zero and so we conclude that $c_k = 0$ for $0 < k < n$, that is $\eta^n \in L_0$. $\square$

We can now proceed to our main result. Let $\theta = (\alpha + \alpha^{-1})/(\alpha^{n-1} + \alpha^{1-n}) \in K$. Note that the conjugates $\sigma^i(\theta)$, $0 \leq i \leq n-1$, are distinct: If $\sigma^i(\theta) = \theta$ then a computation gives $\alpha^{2n} = 1$, which contradicts the fact that $L_0[\alpha] = L$. Hence if $d(x)$ denotes the minimal polynomial of θ over F, then the degree of $d(x)$ is n and $K = F(\theta)$.

THEOREM 3. Let A be any central simple F-algebra of degree n. If the polynomial $d(x)$ (defined above) has a root in A, then $d(x)$ has a conjugate splitting in A.

Proof. If $d(x)$ has a root in A then A contains a (maximal) subfield isomorphic to K, which we will identify with K. Let $\beta \in A \otimes_F L_0$ be as defined above.

We claim there is an element $s \in F^\times$ such that $s\beta + s^{-1}\beta^{-1}$ is invertible in $A \otimes_F L_0$: The extension $L_0[\beta]/L_0$ is a cyclic Galois extension of L_0 (but not necessarily a field). We comput the norm in this extension of the element $s\beta + s^{-1}\beta^{-1}$: $N_{L/L_0}(s\beta + s^{-1}\beta^{-1}) = N_{L/L_0}(s^{-1}\beta^{-1})N_{L/L_0}(s^2\beta^2 + 1) = s^{-2n}b^{-1}((-1)^n - s^{2n}b^2)$. Clearly there is an element $s \in F^\times$ such that this last term is nonzero. For such an s the element $s\beta + s^{-1}\beta^{-1}$ is invertible.

Now let $U = (\alpha + \alpha^{-1})(s\beta + s^{-1}\beta^{-1})^{-1}$ and $V = -(\alpha^{n-1} + \alpha^{1-n})(s\beta + s^{-1}\beta^{-1})^{-1}$. These are elements of $A \otimes_F L_0$ fixed by τ, hence in A. If $c, d \in F$ then $cU + dV = ((c - a^{-1}d)\alpha + (c - ad)\alpha^{-1})(s\beta + s^{-1}\beta^{-1})^{-1}$. If c or d is nonzero then $(c - a^{-1}d)\alpha + (c - ad)\alpha^{-1}$ is a nonzero, hence invertible, element of the field L. By Proposition 2, we have $(cU + dV)^n \in L_0$. Because $U, V \in A$, we obtain $(cU + dV)^n \in F$. Of course if $c = d = 0$ then $(cU + dV)^n = 0 \in F$.

Because our field F is assumed infinite, if we expand $(cU + dV)^n$ we see that there is a unique form $h(x, y) \in F[x, y]$ of degree n such that $h(c, d) = (cU + dV)^n$ for all $c, d \in F$. In particular $h(x, 1) = (xU + V)^n \in A[x]$. Now an easy induction argument shows that for any positive integer k, $(xU + V)^k = U^k \prod_{i=1}^{k}(x + U^{-(k-i)}(U^{-1}V)U^{k-i})$. Since $\theta = -U^{-1}V$, we have $h(x, 1) = \gamma \prod_{i=1}^{n}(x - U^{-(n-i)}\theta U^{n-i})$, where $\gamma = U^n \in F^\times$. Because θ is a root of $h(x, 1)$ and $d(x)$ is the irreducible polynomial of θ over F, we must have $d(x) = h(x, 1)$ and so $d(x)$ has a conjugate splitting in A (with associated constant γ). $\square$

Now let $g(x, y) = y^n d(x/y)$, the homogenization of $d(x)$. For each $\gamma \in F^\times$ let Y_γ denote the curve given by the equation $z^n - \gamma g(x, y)$ and let $F(Y_\gamma)$ denote its function field. Recall that if F_1/F is any field extension, $B(F_1/F)$ denotes the relative Brauer group of the extension, that is the group of Brauer classes of F-central simple algebras split by F_1. We want to compare $B(K/F)$ with the groups $B(F(Y_\gamma)/F)$ as γ varies over the nonzero elements of F. It is well known that any F-central simple algebra split by $F(Y_\gamma)$ must also be split by each of its specializations, that is by each field extension of F obtained by adjoining the coordinates of a point on Y_γ. The field K is a specialization of Y_γ for every γ and so any F-central simple algebra split by $F(Y_\gamma)$ must also be split by K. This is half of the following corollary.

COROLLARY 4. Let K be the field defined above. Then

$$B(K/F) = \bigcup_{\gamma \in F^\times} B(F(Y_\gamma)/F).$$

Proof. One inclusion was shown above. For the opposite inclusion, if the F-central simple algebra A is split by K then A is Brauer equivalent to an algebra B of degree n containing a subfield F–isomorphic to K. Then B contains a root of $d(x)$ and so $d(x)$ has a conjugate splitting in B by the theorem. But then by the equivalence described in the introduction, we see that the class of B lies in $B(F(Y_\gamma)/F)$ for some $\gamma \in F^\times$. $\square$

Notice that this shows that any F–central simple algebra A of degree n split by K must also be split by all of the specializations of Y_γ for some $\gamma \in F^\times$. That is we have produced lots of splitting fields for A. In particular by choosing points on the curve with x and y coordinates in F we produce splitting fields of the form $F[w]$ with $w^n \in F$. Suitable such choices will give cyclic splitting fields of degree n and so cyclic maximal subfields for A, the result of [RS]. However this is not a new proof of this statement, because our method of proof is a generalization of that in [RS].

REFERENCES

L. Childs, *Linearizing of n–ic forms and generalized Clifford algebras* Linear and Multilinear Alg. **5** (1978), 267-278.

D. Haile, *On the Clifford algebra of a binary cubic form*, Amer. J. Math. **106** (1984), 1269-1280.

D. Haile, *On Clifford algebras and conjugate splittings*, J. Algebra **139** (1991), 322–335.

D. Haile, *On Clifford algebras, conjugate splittings, and function fields of curves*, in Ring Theory 1989 in Honor of S.A. Amitsur, L. Rowen ed., Israel Mathematical Conference Proceedings, vol. 1, 1990.

D. Haile and S. Tesser, *On Azumaya algebras arising from Clifford algebras*, J. Algebra **116** (1988), 372–384.

T. Hodges and S. Tesser, *Representing Clifford algebras as crossed-products*, J. Algebra **123** (1989), 500-505.

P. Mammone and J-P. Tignol, *Dihedral algebras are cyclic*, Proc. Amer. Math. Soc. **101** (1987), 217–218.

L. Rowen and D. Saltman, *Dihedral algebras are cyclic*, Proc. Amer. Math. Soc. **84** (1982), 162–164.

S. Tesser, *Representations of a Clifford algebra that are Azumaya algebras and generate the Brauer group*, J. Algebra **119** (1988), 265-281.

J.H.M. Wedderburn, *On division algebras*, Trans. Amer. Math. Soc. **22** (1921), 129–135.

On H-Skew Polynomial Rings and Galois Extensions

SHÛICHI IKEHATA Mathematics Department, Okayama University, Okayama, Japan

GEORGE SZETO Mathematics Department, Bradley University, Peoria, Illinois

1. INTRODUCTION

The notion of Azumaya algebras has been generalized to H-separable extensions of noncommutative rings by Hirata (1968). Many properties of H-separable extensions have been discovered by Ikehata (1981, 1984), Okamoto (1988) and Sugano (1975). In particular, the set of H-separable skew polynomials of automorphism and derivation types were given by one of the present authors (1981 and 1984), and the set of Hopf Galois H-separable polynomials of derivation type were also obtained by Nakajima (1987). The purpose of the present paper is to investigate H-separable skew polynomial rings of finite degree over a center-Galois extension (that is, the center of the coefficient ring is a Galois extension with Galois group generated by the restriction of ρ). We shall characterize an H-separable skew polynomial ring $B[x,\rho]$ over a center-Galois extension B in terms of the commutator subrings of B and the center Z of B in $B[x,\rho]$. A further characterization is also obtained when u ($= x^m$) is in $U(Z^\rho)$ (= the group of units in Z^ρ). Moreover, we shall characterize an H-separable skew polynomial ring of finite degree over an Azumaya algebra B, generalizing Theorem 2.1 in ([I 1]). Furthermore, by using a theorem of DeMeyer (1965), a structure theorem is proved for $B[x,\rho]$ when the Kanzaki hypothesis is satisfied, namely, (1) B is Azumaya

113

over Z and (2) Z is Galois over Z^ρ with Galois group $<\rho/Z>$ of order m isomorphic to $<\rho>$.

The major part of this paper was written when the first author visited the Mathematics Department of Bradley University in 1992. He would like to thank the Mathematics Department for its hospitality.

2. BASIC DEFINITIONS AND NOTATIONS

Let T be a ring extension of A with the same identity, and let C(T) denote the center of T. We call T a separable extension of A if there exist elements $a_i, b_i \in T$, i = 1, ..., k for some integer k, such that $\sum_i a_i b_i = 1$ and $\sum_i t a_i \otimes b_i = \sum_i a_i \otimes b_i t$ for any t in T (where tensor products are over A). If in addition A = C(T), we say T is an Azumaya A-algebra. If $T \otimes T$ is isomorphic to a direct summand of a finite direct sum of copies of T as a T-T-bimodule, we say T is an H-separable extension of A. It is known that any H-separable extension is a separable extension and that any Azumaya algebra is an H-separable extension of its center ([H]).

Throughout this paper, we assume B is a ring with 1 and ρ is an automorphism of B, and we let $B^\rho = \{b \text{ in } B \ / \ \rho(b) = b\}$ denote the fixed subring. We let $<\rho>$ denote the cyclic subgroup of Aut(B) generated by ρ and we let $B[X,\rho]$ be the skew polynomial ring with indeterminate X such that $bX = X\rho(b)$ for any b in B. A monic polynomial f in $B[X,\rho]$ of degree m such that $fB[X, \rho] = B[X,\rho]f$ induces a skew polynomial ring $S = B[X,\rho]/fB[X,\rho]$ which is an extenion ring of B of degree m with $bx = x\rho(b)$ for any b in B where $x = X + fB[X,\rho]$. If S is H-separable over B, we call f an H-separable polynomial. It was shown in Theorem 1.1 in ([I 1]) that f as given above is H-separable if and only if there exist y_i, z_i in S, i = 1, ..., n for some integer n, such that $\sum_i y_i x^k z_i = \delta_{k,m-1}$ for k = 0, 1, ..., m - 1 (where δ is the Kronecker delta) and $by_i = y_i b$ and $\rho^{m-1}(b)z_i = z_i b$ for any b in B. The set $\{y_i, z_i\}$ is called an H-separable system for S over B. It was also shown in Lemma 1 of [I 2] that such an f must have the form $f = X^m - u$ where $u \in U(B^\rho)$ (the group of units of B^ρ) and $\rho^m(b) = u^{-1} bu$ for all b in B. Thus we will assume for the rest of this paper that f has this form, whence $x^m = u$ in S. It follows that x is a unit in S and that ρ has order at most m when restricted to the center of B. We will use Z to denote the center C(B) of B and we will write I_i for the ideal of Z generated by $\{c - \rho^i(c) \ / \ c \in Z\}$; we also set $Ann_B(I_i) = \{a \text{ in } B \ / \ aI_i = 0\}$. Let T be an extension ring of A. We let $V_T(A) = \{t \text{ in } T \mid at = ta \text{ for all } a \text{ in } A\}$ denote the centralizer or commutator of A in T. If G is a finite group of automorphisms of the ring T, we say T is a Galois extension of A with Galois group G if (1) $T^G = A$ and (2) there exist c_i, d_i in T, i = 1, ..., n for some integer n, such that $\sum_i c_i \alpha(d_i) = \delta_{1\alpha}$ (the Kronecker delta) for any α in G. The set $\{c_i, d_i\}$ is called a Galois coordinate system for T. If the center Z of B is Galois over Z^ρ with Galois group $\langle \rho/Z \rangle$ (the restriction of ρ to Z), we call B a center-Galois extension with Galois group $\langle \rho/Z \rangle$. In this case, we note that B is a Galois extension of Z^ρ if $\langle \rho \rangle \cong \langle \rho/Z \rangle$.

3. SKEW POLYNOMIAL RINGS OVER A CENTER-GALOIS EXTENSION

In this section, let $S = B[x,\rho]$ be a skew polynomial ring of finite degree m over a ring B with a free basis $\{1, x, x^2 ..., x^{m-1}\}$ such that $x^m = u$ in $U(B^\rho)$. We shall characterize a center-Galois extension B in terms of the extension S over B and we shall determine some properties of such a B when u is in $U(Z^\rho)$ and when B is Azumaya over Z.

PROPOSITION 3.1

1. $V_S(B) = \{\sum_i x^i b_i \mid \rho^i(b)b_i = b_i b \text{ for all } b \text{ in } B\}$.
2. $V_S(B)$ is commutative if B is commutative.
3. $V_S(Z) = \sum x^i \text{Ann}_B(I_i)$, i = 0, 1, ...,m - 1.
4. If S is H-separable over B, then $C(V_S(B)) = V_S(B)^{\bar{\rho}} = Z$.
5. If S is H-separable over B, then $C(S) = Z^\rho$.

Proof: By noting that $\{1, x, ..., x^{m-1}\}$ is a free basis for S over B, parts 1 and 3 are immediate by the definitions of $V_S(B)$ and $V_S(Z)$.

    2.   Let $\sum_i x^i a_i$ and $\sum_j x^j b_j$ be in $V_S(B)$ and for any b in B, $b(\sum_i x^i a_i) = (\sum_i x^i a_i)b$ and $b(\sum_j x^j a_j) = (\sum_j x^j a_j)b$, so $\rho^i(b)a_i = a_i b$ and $\rho^j(b) b_j = b_j b$. Then, $(\sum_i x^i a_i)(\sum_j x^j b_j) = \sum_{i,j} x^{i+j}\rho^j(a_i)b_j = \sum_{i,j} x^{i+j} b_j a_i$ and $(\sum_j x^j b_j)(\sum_i x^i a_i) = \sum_{i,j} x^{j+i}\rho^i(b_j)a_i = \sum_{i,j} x^{i+j} a_i b_j$. Since B is commutative, $a_i b_j = b_j a_i$. Thus $V_S(B)$ is commutative.

4.   Since S is an H-separable extension of B with a free basis, $B = V_S(V_S(B))$ ([S], (5), P. 296). Hence, $C(V_S(B)) = V_S(V_S(B)) \cap V_S(B) = B \cap V_S(B) = V_B(B) = Z$.

5. $C(S) = V_S(S) = V_S(B, x) = V_Z(x) = Z^\rho$ by part 4.

THEOREM 3.2 S is an H-separable extension of B if and only if $V_S(B)$ is a Galois extension of Z^ρ with Galois group $<\bar{\rho}/V_S(B)>$.

Proof: Let $f(X)$ be an H-separable polynomial. Then, by Theorem 1.1 in [I 1], there exist $\{y_i, z_i$ in S, $i = 1, \ldots, k$ for some integer $k\}$ such that $by_i = y_i b$, $\rho^{m-1}(b)z_i = z_i b$ for any b in B, $\sum_i y_i x^{m-1} z_i = 1$ and $\sum_i y_i x^k z_i = 0$ for each $k = 0, 1, \ldots, m-2$. We know that $x^{m-1} z_i$ is in $V_S(B)$ and $\sum_i y_i \bar{\rho}^{j}(x^{m-1} z_i) = \sum_i y_i x^{m-1} \bar{\rho}^{j}(z_i) = \sum_i y_i x^{m-1-j} z_i x^j = 0$ for each $j = 1, \ldots, m-1$. Since $f(X)$ is an H-separable polynomial, $f(X) = X^m - u$ for some u in $U(B^\rho)$ and $\rho^m(b) = u^{-1} bu$ for any b in B ([I 2], Lemma 1). Thus, the order of $\bar{\rho}/V_S(B)$ is m. Moreover, we claim that $(V_S(B))^{\bar{\rho}} = Z^\rho$. In fact, for any $y = \sum_{j=0}^{m-1} x^j d_j$ in $(V_S(B))^{\bar{\rho}}$, $\bar{\rho}(y) = y$, so we have that $\rho(d_j) = d_j$ for each j. Also, $by = yb$ for any b in B, so $\rho^j(b)d_j = d_j b$. Hence $d_j = 0$ for each $j = 1, \ldots, m-1$ ([I 2], Lemma 1); and so d_i is in Z^ρ. Noting that $Z^\rho \subset (V_S(B))^{\bar{\rho}}$ we conclude that $(V_S(B))^{\bar{\rho}} = Z^\rho$. Thus $V_S(B)$ is Galois over Z^ρ with Galois coordinate system $\{y_i, x^{m-1} z_i\}$.

    Conversely, since $V_S(B)$ is Galois over Z^ρ with Galois group $<\bar{\rho}/V_S(B)>$ of order m, there exist Galois coordinate system $\{y_i, \beta_i$ in $V_S(B)$; $i = 1, \ldots, k$ for some integer $k\}$ such that $\sum_i y_i \bar{\rho}^{j}(\beta_i) = \delta_{1_\rho j}$. Noting that x is a unit in S, we put $z_i = x^{-(m-1)} \beta_i$. Then $\rho^{m-1}(b)z_i = \rho^{m-1}(b)x^{-(m-1)}\beta_i = x^{-(m-1)}b\beta_i = x^{-(m-1)}\beta_i b = z_i b$ for any b in B. Therefore, we have a set $\{y_i, z_i\}$ such that $\sum_i y_i x^{m-1} z_i = \sum_i y_i \beta_i = 1$ and $\sum_i y_i x^k z_i = \sum_i y_i x^{k-(m-1)}\beta_i = \sum_i y_i \bar{\rho}^{m-1-k}(\beta_i)x^{k-(m-1)} = 0$ for each $k = 0, 1, \ldots, m-2$.

Thus $f(X)$ is an H-separable polynomial ([I 1], Theorem 1.1).

    Next are characterizations of a center-Galois extension.

THEOREM 3.3 Let I_i be the ideal of Z generated by $\{c - \rho^i(c) / c\} \in Z$. The following are equivalent:

1. Z is Galois over Z^ρ with Galois group $<\rho/Z>$ of order m (that is, B is center-Galois).
2. S is an H-separable extension over B and $V_S(Z) = B$.
3. S is an H-separable over B and $V_S(B) = Z$.
4. S is an H-separable extension over B and $Ann_B(I_i) = \{0\}$.

Proof: $1 \rightarrow 3$. Since Z is Galois over Z^ρ with Galois group $<\rho/Z>$ of order m, S is H-separable over B by the proof of the sufficiency of Theorem 3.2; that is, let $\{\alpha, \beta_i \text{ in } Z; i = 1, ..., k \text{ for some integer } k\}$ be a Galois coordinate system. Then we can show that $\{\alpha_i, x^{-(m-i)}\beta_i\}$ is an H-separable system for S. Hence $V_S(B)$ is Galois over Z^ρ with Galois group $<\rho/V_S(B)>$ of order m by Theorem 3.2. Since $Z \subset V_S(B)$, both Z and $V_S(B)$ are Galois over Z^ρ with Galois groups induced by ρ and of the same order m, $Z = V_S(B)$.

$3 \rightarrow 1$ is a consequence of Theorem 3.2 again.

$2 \rightarrow 3$. Since $V_S(B) \subset V_S(Z)$ which is B by hypothesis, $V_S(B) \subset Z$. Thus $V_S(B) = Z$ (for $Z \subset V_S(B)$ clearly).

$3 \rightarrow 2$. Since $V_S(B) = Z$, $V_S(V_S(B)) = V_S(Z)$. But B is an B-direct summand of S and S is H-separable over B, so $B = V_S(V_S(B))$ ([S], (5), P. 296). Hence, $B = V_S(Z)$.

$2 \leftrightarrow 4$. Noting that $V_S(Z) = \sum_{i=0}^{m-1} Ann_B(I_i)x^i$ by Proposition 3.1 (3), we have that $V_S(Z) = B$ if and only if $Ann_B(I_i) = \{0\}$ for each $i = 1, ..., m - 1$.

We shall show a characterization of a center-Galois extension B when u is in $U(Z^\rho)$.

THEOREM 3.4 The following conditions are equivalent:

(a) Z is Galois over Z^ρ with Galois group $<\rho/Z>$ of order m and $u \in U(Z^\rho)$.

(b) (i) S is an H-separable extension of B, (ii) $Ann_Z(I_i) = \{0\}$ for each $i = 1, ..., m-1$, and (iii) $B = B^\rho Z$.

Proof: Let Z be Galois over Z^ρ with Galois group $<\rho/Z>$ of order m. Then (i) and (ii) hold by Theorem 3.3 $(1) \to (4)$. For (iii), it can be checked that $\{tr(\alpha_i \_), \beta_i\}$ is a projective dual basis for B over B^ρ where $\{ \alpha_i, \beta_i$ in Z, i = 1, ..., k for some integer k$\}$ is a Galois coordinate system for Z and $tr(b) = \sum_{j=0}^{m-1} \rho^j(b)$ for each b in B.

Thus $b = \sum_i tr(b\alpha_i)\beta_i$ which is in $B^\rho Z$; and so $B = B^\rho Z$.

Conversely, it suffices by (3) of Theorem 3.3 to show that $V_S(B) = Z$. In fact, let $\sum_i x^i a_i$ be an element in $V_S(B)$; then, for any b in B^ρ, $b\left(\sum_i x^i a_i\right) = \left(\sum_i x^i a_i\right) b$.

Thus $\rho^i(b)a_i = ba_i = a_i b$. Since $B = B^\rho Z$, a_i is in Z for each i. Also, for any b in Z, $b\left(\sum_i x^i a_i\right) = \left(\sum_i x^i a_i\right) b$, so $\rho^i(b)a_i = a_i b$. Hence $a_i(b - \rho^i(b)) = 0$ for each i and b in Z. But $Ann_Z(I_i) = \{0\}$ for any i = 1, ..., m-1, so $a_i = 0$ for any i = 1, ..., m-1. Therefore $V_S(B) = Z$ (for $Z \subset V_S(B)$ is clear).

4 SKEW POLYNOMIAL RINGS OVER AN AZUMAYA ALGEBRA

In this section, we shall give a sufficient and necessary condition for S being an H-separable extension over an Azumaya algebra B. Our result generalizes Theorem 2.1 in ([I 2]) for S over a commutative ring B. Then a structure theorem for S over B satisfying the Kanzaki hypothesis is derived.

THEOREM 4.1   S is an H-separable extension over an Azumaya algebra B if and only if S is an Azumaya Z^ρ-algebra.

Proof: Let S be an Azumaya Z^ρ algebra. Since S is a free left module over B, S is an H-separable extension over B ([I 1], Theorem 1). But then $B = V_S(V_S(B))$ ([S], (5), P. 296). Again, S is an H-separable extension over B so $V_S(B)$ is Galois over Z^ρ by Theorem 3.2. Hence $V_S(B)$ is a separable subalgebra of S over Z^ρ.

Thus $B = V_S(V_S(B))$ is also a separable subalgebra of S over Z^ρ by the commutator theorem for Azumaya algebras ([DI], Theorem 4.3, P. 57). Therefore B is Azumaya over Z ([DI], Theorem 3.8, P. 55).

   Conversely, Since S is an H-separable extension over B and B is an Azumaya Z-algebra, S is also an Azumaya algebra over its center ([O], Theorem 1). By Proposition 3.1, the center of S is Z^ρ, so S is Azumaya over Z^ρ.

   We recall that a ring B together with a finite automorphism group G satisfies the Kanzaki hypothesis if (1) B is Azumaya over Z and (2) Z is Galois over Z^G with Galois group G/Z induced by and isomorphic to G ([D 1], [SW], [SM]). Now we consider the Kanzaki hypothesis in the case G = <ρ>. We note that when B satisfies the hypothesis, S is an H-separable extension of B.

COROLLARY 4.2 If (B,<ρ>) satisfies the Kanzaki hypothesis, then S, B^ρ and $Z[x,\rho]$ are Azumaya Z^ρ-algebras and $S \cong B^\rho \otimes_{Z^\rho} Z[x,\rho]$.

Proof: Since B satisfies the Kanzaki hypothesis, $B \cong B^\rho \otimes_{Z^\rho} Z$ and B^ρ is Azumaya over Z^ρ ([D 1], Lemma 2, P. 119). Hence $S = B[x,\rho] \cong B^\rho \otimes_{Z^\rho} Z[x,\rho]$. By Theorem 4.1, S is an Azumaya Z^ρ algebra, so B^ρ and $Z[x,\rho]$ are Azumaya Z^ρ-algebras ([DI], Theorem 4.4, P. 58).

   A similar structure theorem for the skew polynomial subrings $B[x^i,\rho^i]$ of S for each i = 1, ..., m-1 can be proved.

THEOREM 4.3 If (B,<ρ>) satisfies the Kanzaki hypothesis, then (1) for each i = 1, ..., m-1, B^{ρ^i} is Azumaya over Z^{ρ^i}, and (2) $B[x^i,\rho^i] \cong B^{\rho^i} \otimes_{Z^{\rho^i}} Z[x^i,\rho^i]$ is Azumaya over Z^{ρ^i}, where $\otimes$ is over Z^{ρ^i}.

Proof: 1. For i = 1, B^ρ is Azumaya over Z^ρ ([D 1], Lemma 2, P.119). For i>1, since $B = B^\rho Z \cong B^\rho \otimes_{Z^\rho} Z$ ([D 1], Lemma 2, P. 119), $B^{\rho^i} = (B^\rho Z)^{\rho^i}$

$\cong (B^\rho \otimes_{Z^\rho} Z)^{1 \otimes \rho^i} = B^\rho \otimes_{Z^\rho} Z^{\rho^i}$; and so B^{ρ^i} is an Azumaya Z^{ρ^i}-algebra.

   2. By (1), $B[x^i,\rho^i] \cong (B^\rho \otimes_{Z^\rho} Z)[x^i,\rho^i] \cong (B^\rho \otimes_{Z^\rho} Z^{\rho^i} \otimes_{Z^{\rho^i}} Z)[x^i,\rho^i]$

$\cong (B^{\rho^i} \otimes_{Z^{\rho^i}} Z)[x^i,\rho^i] = B^{\rho^i} \otimes_{Z^{\rho^i}} Z[x^i,\rho^i]$. Since Z is Galois over Z^ρ with Galois group $<\rho / Z>$, $Z[x^i, \rho^i]$ is Azumaya over $Z^{\rho'}$. ([I 1], Theorem 2.2, P. 23). Thus $B^{\rho'} \otimes_{Z^{\rho'}} Z[x^i, \rho^i] \cong B[x^i, \rho^i]$ is an isomorphism as Azumaya $Z^{\rho'}$-algebras.

Proposition 3.1 shows that the center of $V_S(B)$ is Z if S is an H-separable extension over B and Theorem 3.2 states that $V_S(B)$ is Galois over Z^ρ with Galois group $<\bar{\rho}/V_S(B)>$ of order m if and only if S is an H-separable extension of B. Thus we derive the following proposition for $V_S(B)$.

PROPOSITION 4.4 If S is an H-separable extension of B, then $V_S(B)$ is an Azumaya Z-algebra.

Proof: By Theorem 3.2, $V_S(B)$ is Galois over Z^ρ so $V_S(B)$ is separable over Z^ρ. But the center of $V_S(B)$ is Z by Proposition 3.1, so $V_S(B)$ is Azumaya over Z.

REFERENCES

[D 1] DeMeyer, F. R. (1965). Some Notes on the General Galois Theory of Rings,   Osaka J. Math., 2: 117-127.

[D 2] DeMeyer, F. R. (1966). Galois Theory in Separable Algebras over Commutative Rings, Illinois J. Math., 10: 287-295.

[DI] DeMeyer, F. R. and Ingraham, E. (1971). Separable algebras over commutative rings, Lecture Notes in Mathematics, 181: Springer-Verlag, Berlin-Heidelberg-New York.

[H]   Hirata, K. (1968). Some Types of Separable Extensions, Nagoya Math J., 33: 107-115.

[I 1] Ikehata, S. (1981). Azumaya Algebras and Skew Polynomial Rings, Math J. Okayama Univ., 23: 19-32.

[I 2] Ikehata, S. (1984). Azumaya Algebras and Skew Polynomial Rings, II, Math J. Okayama Univ., 26: 49-57.

[I 3] Ikehata, S. (1981). Note on Azumaya Algebras and H-Separable Extensions, Math J. Okayama Univ., 23: 17-18.

[N]   Nakajima, A. (1987). P-Polynomials and H-Galois Extensions, J. Alg., 110: 124-133.

[O]   Okamoto, H. (1988). On Projective H-Separable Extensions of Azumaya Algebras, Results in Math., 14: 330-332.

[S]   Sugano, K. 1975). On a Special Type of Galois Extensions, <u>Hokkaido Math J.</u> , <u>4</u>: 123-128.

[SW] Szeto, G. and Wong, Y.F. (1983). On Separable Noncyclic Extensions of Rings, <u>J. Austral. Math Soc. Series A</u>, <u>34</u>: 394-398.

[SM] Szeto, G. and Ma, L.J. (1988). On Rings Whose Center Are Galois Extensions, <u>Portugaliae Math.</u>, <u>45</u>: 75-82.

Separability and the Jones Polynomial

LARS KADISON   Department of Mathematics
and Physics, Roskilde University, 4000 Roskilde, Denmark

1   Introduction

Jones' index theory of type II_1 von Neumann algebra subfactors was published in 1983, and led in the spring of the following year to a new polynomial invariant of knots and links. Subsequently, the Jones polynomial was generalized in different directions and several old problems of Tait's in knot theory were solved. Certain key ingredients of Jones' theory may be reduced to algebra in different ways. For example, the Jones polynomial may be defined from certain traces of Ocneanu's on a sequence of finite dimensional algebras named after Hecke. A second example: the semi-discrete index spectrum of II_1 subfactors may be obtained from the classification of matrix norms of the 0-1 matrices - accomplished long ago - the 0-1 matrices arising as the inclusion matrices of the multimatrix ϵ_i-algebras $\mathcal{A}_{\beta,n} \subseteq \mathcal{A}_{\beta,n+1}$. Another algebraic direction to Jones' theory was started by M. Pimsner and S. Popa in [16] in which they find an "orthonormal basis" that shows a II_1 factor M to be a finitely generated projective module over a finite index subfactor N, the index being the Hattori-Stallings rank of the module [10].

In pursuing the algebraic direction of Pimsner and Popa, the author and D. Kastler in [10] showed that $N \subseteq M$ is a separable extension of rings. The reader will recall the notion of a separable field extension [11] and separable algebra over a field [15], which received various insightful generalizations in Hochschild's homological algebra [7, 1945], and in Auslander and Goldman's theory of the Brauer group of a commutative ring [1, 1960]. In this paper we build from a small system of axioms, one being relative separability and another its module-theoretic dual notion of split extension, the Jones theory leading up to the V_L polynomial for a link L. We define finite separable extensions of k-algebras, and show that these possess the key elements of

Jones' theory: basic construction, index, and, upon iteration of the basic construction to produce a tower of algebras, a countable family of idempotents satisfying braid-like relations. Including trace in this picture, one trace T on the tower of algebras extending the previous by composing it with the conditional expectation, the Jones polynomial V_L of a link L can then be defined by normalizing the trace of a representation of the braid group in a finite separable T-extension.

The structure theory of finite separable extensions of algebras, its relations with representation theory of groups, and the duality of separable and split extension is treated in my paper [9]. In the present paper we further discuss a dimension question and what relation finite separability has with quasi-Frobenius extensions.

2   Finite separable extensions

Let k be a commutative ring, and k° its group of units. Let S be a subalgebra of a faithful k-algebra A such that $1_A \in S$. We identify k with $k1_A$. We consider only natural module and bimodule structure coming from inclusion $S \hookrightarrow A$ and tensor product over S. Let μ denote the multiplication map $A \otimes_S A \to A$, an A-A bimodule morphism defined by $a_0 \otimes a_1 \mapsto a_0 a_1$.

Definition 2.1 A is called a finite separable extension of S if there exists an element $f \in A \otimes_S A$, an S-S bimodule homomorphism $E : A \to S$, and $\tau \in k^\circ$ such that

1. $af = fa \quad (\forall a \in A)$ and $\mu(f) = 1$;

2. $E(1) = 1$;

3. $\mu(1 \otimes_S E)f = \mu(E \otimes_S 1)f = \tau$.

An element f satisfying axiom 1 is called a separating element, or a separability element, and its existence alone defines a separable extension of rings, a theory generalizing separable algebras and developed by Sugano and several others [3].

The existence of f is equivalent to μ being a split epimorphism of A-bimodules, which is in turn equivalent to the vanishing of relative Hochschild cohomology groups [1] with arbitrary coefficients $(n > 0)$,

$$H^n(A, S; -) = 0.$$

The map $E : A \to S$ satisfying axiom 2 is a <u>conditional expectation</u> as in operator theory, and its existence for a subalgebra $S \subseteq A$ defines a split extension of rings. It is equivalent to requiring the subalgebra S be a direct summand in the bimodule $_SA_S$: since the inclusion map splits the kernel exact sequence of the S-S epimorphism $E : A \to S$, we note that

$$A = S \oplus \ker E.$$

Conditional expectations and separating elements are not unique, but axiom 3 demands the existence of a conditional expectation $E : A \to S$ and separating element in $A \otimes_S A$

$$f = \tau \sum_{i=1}^{n} x_i \otimes_S y_i$$

such that $\tau \in k^\circ$ and

(1)
$$\sum_{i=1}^{n} E(x_i)y_i = \sum_{i=1}^{n} x_i E(y_i) = 1.$$

Indeed, a short computation reveals that we may choose $x_1 = y_1 = 1$ and $x_i, y_i \in \ker E$ for $i = 2, \ldots, n$. We will say that E and f are <u>compatible</u> in case they satisfy axiom 3. Now fix the notations $S \subseteq A$, f, E, x_i, y_i, μ, and τ for the rest of this paper.

Lemma 2.1 For every $a \in A$, we have

(2)
$$\sum_{i=1}^{n} E(ax_i)y_i = \sum_{i=1}^{n} x_i E(y_i a) = a.$$

Proof. One can define a linear map from $A \otimes_S A \otimes_k \hom_S(A, S) \to \hom_k(A, S)$ and make use of axiom 1 together with equation (1). The argument is repeatable on the right. $\square$

Corollary 2.1 E is a nondegenerate S-valued bilinear form on A such that $\{E(-x_i)\}_{i=1}^{n}$ is a dual basis of $\{y_i\}_{i=1}^{n}$ for the projective module $_SA$ and $\{E(y_i-)\}_{i=1}^{n}$ is a dual basis of $\{x_i\}_{i=1}^{n}$ for A_S.

Proof. Nondegeneracy of E follows from assuming $E(ax) = 0$ for every $x \in A$, then $a = \sum E(ax_i)y_i = 0$. $\square$

---

[1] defined in [8]

3 The basic construction

Define a k-algebra structure on the k-module, $A \otimes_S A$, on which we place a k-algebra structure with multiplication given by

$$(a_0 \otimes_S a_1)(a_2 \otimes_S a_3) = a_0 E(a_1 a_2) \otimes_S a_3.$$

The unity element of A_1 is $1 = \sum_{i=1}^n x_i \otimes_S y_i$.

Define <u>index</u> of S in A, $[A:S]_E = \tau^{-1}$, an invertible element in k. This definition is independent of f since $\mu(1) = \tau^{-1}$.

Proposition 3.1 A_1 is a finite separable extension of A with index $[A:S]_E$.

Proof. A separating element

$$f = \sum_{i=1}^n x_i \otimes_S 1 \otimes_S y_i$$

is compatible with the conditional expectation

$$E_1 = \tau\mu : A_1 \to A \quad \square$$

Remark 3.1 Note that the element $e_1 = 1 \otimes_S 1$ in A_1 is an idempotent, and a cyclic generator of A_1 as an A-A bimodule. Also note the identities,

$$(3) \qquad\qquad e_1 a e_1 = E(a) e_1,$$

for every $a \in A$ and

$$(4) \qquad\qquad E_1(e_1) = \tau 1.$$

Proposition 3.2 If B is a finite separable extension of A, which in turn is a finite separable extension of S, with conditional expectations E_1 and E_2, resp., then B is a finite separable extension of S with index satisfying Lagrange's equation,

$$[B:S]_{E_2 \circ E_1} = [B:A]_{E_1} [A:S]_{E_2}.$$

Proof. Let $(B, A, E_1, f_1 = \tau_1 \sum_{i=1}^{m} u_i \otimes_A v_i)$, $(A, S, E_2, f_2 = \tau_2 \sum_{i=1}^{n} x_i \otimes_S y_j)$ be the data for finite separable extension. Note that $E = E_2 \circ E_1 : B \to S$ is a conditional expectation, and

$$f = \tau_1 \tau_2 \sum_{i=1}^{m} \sum_{j=1}^{n} u_i x_j \otimes_S y_j v_i$$

is a compatible separating element in $B \otimes_S B$. $\square$

As a corollary we note that finite separable extension is closed under tensor product with index behaving multiplicatively. Since every algebra is a finite separable extension of itself with index 1, we are permitted a change of ring k on a finite separable extension of k-algebras - with no alteration to the index.

4 Examples

1. **Matrices.** The full matrix algebra $M_n(A)$ over any k-algebra A is a finite separable extension of A (embedded in the constant diagonals) so long as n is invertible in k. Of the n separability elements, $(j = 1, 2, \ldots, n)$

$$f_j = \sum_{i=1}^{n} E_{ij} \otimes_A E_{ji},$$

(E_{ij} is the (i, j)-matrix unit) we average to obtain $f = \frac{1}{n} \sum_{j=1}^{n} f_j$ as our separating element. The conditional expectation

$$E(X) = \frac{1}{n} \sum_{i=1}^{n} X_{ii} \quad \text{where} \quad X = (X_{ij}) \in M_n(A).$$

is easily computed to be compatible with f, having index reciprocal $\tau = \frac{1}{n^2}$.

Taking different weighted averages of the elements f_j we can find separating elements and compatible expectations with index other than n^2.

2. **Subfactors.** Let N be a subfactor of M with Jones index $n \leq [M : N] < n + 1$ If $\{m_j\}_{j=1}^{n+1}$ is the Pimsner-Popa orthonormal basis [16] with respect to

the trace-preserving conditional expectation $E : M \to N$, a separating element compatible with E is then given by

$$\frac{1}{[M : N]} \sum_{j=1}^{n+1} m_j \otimes m_j^*$$

This example is observed and proven by D. Kastler and the author in [10]. The proof of these assertions follows from the relations in [16, p. 65], the isomorphism in [4, p. 189], the relations $e_1 x e_1 = E(x) e_1$ and the implication $e_1 x = e_1 y \Rightarrow x = y$, where e_1 denotes the projection of $L^2(M, \mathrm{tr})$ onto $L^2(N, \mathrm{tr})$.

3. **Finite Separable Extensions of Fields** F_2/F_1 with characteristic coprime to the degree n. Let α be a primitive element, $F_2 = F_1(\alpha)$, with minimal polynomial

$$p(x) = x^n - \sum_{i=0}^{n-1} c_i x^i$$

Let

$$E = \frac{1}{n} \mathrm{trace} : F_2 \to F_1,$$

the normalized trace, where trace is a nondegenerate bilinear form on the F_1-vector space F_2 with dual bases [12, cf. p. 213] $\{\alpha^i\}_{i=0}^{n-1}$ and

$$\{ \frac{\sum_{j=0}^{i} c_j \alpha^j}{p'(\alpha) \alpha^{i+1}} \}_{i=0}^{n-1}.$$

A separating element is given [14] by

$$f = \sum_{i=0}^{n-1} \alpha^i \otimes_{F_1} \frac{\sum_{j=0}^{i} c_j \alpha^j}{p'(\alpha) \alpha^{i+1}}.$$

Denoting f by $\sum_{i=0}^{n-1} u_i \otimes v_i$ where $E(u_i v_j) = \frac{1}{n} \delta_{i,j}$, we easily compute $\sum u_i E(v_i) = \sum u_i E(u_0 v_i) = \frac{1}{n}$, since $u_0 = 1$. Letting $1 = \sum b_i v_i$, we get $\sum E(u_i) v_i = \sum b_j E(u_i v_j) v_i = \frac{1}{n} \sum b_j v_j = \frac{1}{n}$. Hence, f and E are compatible with index n. In characteristic p the index is n (mod p).

4. **Crossed product algebras.** Let H be a subgroup of G with finite index $[G : H] \in k^\circ$, B a k-algebra with action $\alpha : G \to \text{Aut } B$. Then $A = B \times_\alpha G$ is a finite separable extension of $S = B \times_\alpha H$. For if $\{g_i\}_{i=1}^n$ is a left transversal of H in G, then

$$f = \frac{1}{[G : H]} \sum_{i=1}^n g_i \otimes_S g_i^{-1}$$

is a separating element compatible with the natural projection

$$\pi_H : B \times_\alpha G \to B \times_\alpha H;$$

whence $\tau = \frac{1}{[G:H]}$. Note that group algebras, and specifically those generated by Sylow p-subgroups of finite groups over characteristic p are included in this example. The proof of these assertions is elementary.

Remark 4.1 Galois extensions of commutative rings, multimatrix extensions $M_{n_1}(S) \times \cdots \times M_{n_t}(S)$, are finite separable extensions as are separable algebras over a local or global field (if dimension is coprime to the characteristic).

5 Tower of algebras

We have seen in section 2 that the basic construction is itself a finite separable extension with canonical expectation and same index. Let A_{i+1} be defined inductively for i = 1,2,3,... as the basic construction of the finite separable extension $A_{i-1} \subseteq A_i$. We make use of the natural notation $A_0 = A$, multiplication map $\mu_{i+1} : A_i \otimes_{A_{i-1}} A_i \to A_i$, and conditional expectation $E_{i+1} = \tau \mu_{i+1}$. By this iteration of the basic construction a tower of algebras over S is generated:

$$S \subseteq A \subseteq A_1 \subseteq A_2 \subseteq \cdots$$

Theorem 5.1 If the ground ring k of a finite separable extension $S \subset A$ possesses an invertible solution t to the quadratic equation $(t + 1)^2 \tau = t$, then for every n there exists a nontrivial homomorphism of the braid group B_n into the group of units in A_{n-1}. Under the same hypothesis, there exists a nontrivial homomorphism of the Hecke algebra $H(t, n)$ into A_{n-1}.

Proof. We have the idempotent $e_{i+1} = 1 \otimes_{A_{i-1}} 1$ in A_{i+1}; the family of idempotents $\{e_i\}_{i=1}^{\infty}$ in the tower of algebras $S \subseteq A \subseteq A_1 \subseteq \ldots$ satisfies the braid-like relations ($|i - j| \geq 2$):

(5) $$e_i e_{i+1} e_i = \tau e_i,$$

(6) $$e_{i+1} e_i e_{i+1} = \tau e_{i+1},$$

(7) $$e_i e_j = e_j e_i.$$

Equation (6) follows from equations (3) and (4). Equation (7) is a simple consequence of noting that S is the centralizer subalgebra of e_1 in the basic construction. Equation (5) is the tedious computation that $e_1 e_2 e_1 = \tau e_1$.

Map the Artin generators $\{\sigma_i : i = 1, \ldots, n-1\}$ of B_n as follows:

(8) $$\Phi_n : \quad \sigma_i \longmapsto w_i = (t+1)e_i - 1.$$

One can readily check that the w_i are units of A_{n-1} and satisfy the Artin relations:

$$\sigma_i \sigma_j = \sigma_j \sigma_i$$

$$\sigma_{i+1} \sigma_i \sigma_{i+1} = \sigma_i \sigma_{i+1} \sigma_i$$

Hence, the map Φ_n extends multiplicatively to a homomorphism of B_n into the group of units in A_{n-1}.

The Hecke algebras $H(t, n)$ have the standard presentation

$$\langle g_1, \ldots, g_{n-1} | g_i^2 = (t-1)g_i + t, \ \ g_i g_{i+1} g_i = g_{i+1} g_i g_{i+1}, g_i g_j = g_j g_i, \ \forall i, j : |i-j| \geq 2 \rangle.$$

Since the w_i also satisfy $w_i^2 = (t-1)w_i + t$, an algebra homomorphism is obtained by sending each g_i onto w_i. $\square$

6   Markov traces

Recall that a <u>trace</u> on a k-algebra B is a k-linear map $T : B \to k$ satisfying for each $x, y \in B$, $T(xy) = T(yx)$. In addition, we assume a trace to be normalized: $T(1) = 1$.

Definition 6.1 Suppose there exists a trace T_S on S. We say A is a <u>finite separable</u> <u>T-extension</u> of S if A is a finite separable extension of S with conditional expectation $E : A \to S$ such that $T_S \circ E$ defines a trace on A we denote by T_A.

Remark 6.1 Note that T_A restricted to S equals the trace T_S, We call T_A the Markov trace over S. Note the identity ($\forall s \in S, \quad a \in A$)

(9) $$T_A(sa) = T_S(sE(a))$$

Examples of finite separable T-extensions abound: each of the four main examples above of finite separable extension is also a T-extension with respect to the canonical trace on S.

Recall the notation $e_1 = 1 \otimes_S 1$ and $E_1 = \tau\mu$.

Theorem 6.1 Suppose A is a finite separable T-extension of S. Then the basic construction A_1 is a finite separable T-extension of A.

Proof. We must check that $T_{A_1} = T_A \circ E_1$ defines a trace. Note that the k-linear map $T_{A_1} : A_1 \to k$ is given by

$$T_{A_1}\left(\sum_{i=1}^{N} a_i e_1 b_i\right) = \tau \sum_{i=1}^{N} T_A(a_i b_i).$$

A computation with simple tensors and using equation (7) shows that T_{A_1} is a trace:

$$T_{A_1}(a_0 e_1 a_1 a_2 e_1 a_3) = T_{A_1}(a_0 E(a_1 a_2) e_1 a_3) =$$
$$\tau T_A(E(a_1 a_2) a_3 a_0) = \tau T_S(E(a_1 a_2) E(a_3 a_0)) =$$
$$T_{A_1}(a_2 e_1 a_3 a_0 e_1 a_1)$$

This together with proposition 1.1 completes the proof. $\square$

We note the identity, which iterates up the tower of algebras over A ($\forall a \in A$):

(10) $$T_{A_1}(ae_1) = \tau T_A(a)$$

Abbreviating the n'th Markov trace T_{A_n} to T_n, we get values for trace such as $T_5(e_4 e_3 e_2 e_3 e_5) = \tau^4$. In the next section we will need only know the values of trace on the e_i-algebra $\mathcal{A}_{\beta,n}$ generated by $1, e_1, \ldots, e_n$ within the tower ($\beta = [A : S]$).

7   The Jones polynomial

Jones associates in [5] a Laurent polynomial in $t^{\frac{1}{2}}$ to an oriented link L. This has been done in various ways via von Neumann algebras and Hecke algebras. In this section we show how it may be done with finite separable T-extensions.

Given a link L, we can find at least one braid α on n strings, which when closed up gives back the link L (a theorem of Alexander). Illustrations of closure are given below, and further reading on knot theory can be found in [2] and [6]. Now there is uncertainty in the correspondence $L \rightsquigarrow \alpha$. It turns out that for any other braid β on n strings, $\beta\alpha\beta^{-1}$ as well as the two braids $\sigma_n^{\pm 1}\alpha$ in B_{n+1} all [2] close up to give L: moreover, by a theorem of Markov the various finite compositions of conjugation and left multiplication of $\sigma_n^{\pm 1}, \sigma_{n+1}^{\pm 1}, \ldots$ are all the braids closing up to give L: these conjugation and left multiplication operations are called the first and second Markov moves, respectively.

Consider a finite separable T-extension A of S with index reciprocal $\tau = \frac{t}{(t+1)^2}$, such as may be obtained from a change of rings,

$$k \rightsquigarrow k[t, t^{-1}]/((t+1)^2 \tau - t).$$

By theorem 5.1 there is a group homomorphism $\Phi_n : B_n \to A_{n-1}^{\,\circ}$ (cf. equation (8)). Making use of trace, we note that $T \circ \Phi_n$ is a character function on B_n (i.e., a scalar-valued function constant on conjugacy classes of the group). We need an adjustment or normalization for the other possible types of braids closing up to give the same link: the Jones polynomial $V_L(t)$ in $k[\sqrt{t}, \frac{1}{\sqrt{t}}]$ gets the job done.

Let exponent $e(\alpha)$ equal the sum of powers in α written as a word in the σ_i. This is an invariant of word reduction in B_n! Define the Jones polynomial of the link L with corresponding braid α in B_n as follows:

$$(11) \qquad V_L(t) = (-\frac{t+1}{\sqrt{t}})^{n-1}(\sqrt{t})^{e(\alpha)} T(\Phi_n(\alpha))$$

To prove this an invariant under the second Markov move, note that $T(\Phi_{n+1}(\sigma^{\pm 1}\alpha)) = T((t^{\pm 1}+1)e_n \Phi_n(\alpha)) - T(\Phi_n(\alpha)) = -\frac{t^{0,1}}{t+1}T(\Phi_n(\alpha))$; plugging

---

[2] The Artin generators σ_i cross the i'th and i+1'st strings while fixing the other strings.

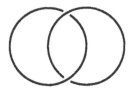

Figure 1. σ_1^2. The Hopf link

this in the right-hand side above, with appropriate changes made to the powers $n - 1$ and $e(\alpha)$, one sees the polynomial does not change. It follows from Markov's and Alexander's theorems that $V_L(t)$ is an isotopy invariant of links.

Example 1: *The Hopf link.* The braid σ_1^2 in B_2 closes up to give the Hopf link, which is two interlocking circles oriented as shown above. The exponent of this braid is 2 and $n = 2$. Hence,

$$V_L(t) = -\frac{t+1}{\sqrt{t}}tT((e_1(t+1)-1)^2) =$$

$$-(t+1)\sqrt{t}(t - \frac{2t}{t+1} + 1) = -t^{\frac{5}{2}} - t^{\frac{1}{2}}.$$

Example 2: *The left-handed trefoil knot.* The braid σ_1^3 in B_2 closes up to give the left-handed trefoil knot as shown below. The equality sign below should be understood as an isotopy in the ambient space. We compute:

$$V_L(t) = (-\frac{t+1}{\sqrt{t}})\sqrt{t}^3 T((t^3+1)e_1 - 1) =$$

$$-\frac{t}{t+1}((t^3+1)t - (t+1)^2) = -t^4 + t^3 + t.$$

 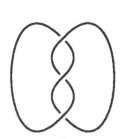 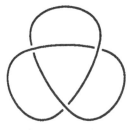

Figure 2. σ_1^3. Its closure = the left-handed trefoil knot.

The mirror image link. Suppose the braid $\alpha = \sigma_1^{n_1} \cdots \sigma_k^{n_k}$ closes up to give the oriented link L. Now we hold a mirror below the link as we look down on it, which turns overpasses to underpasses as we follow the arrows around the link (or knot): this is the mirror image knot denoted by L^-. Clearly, the braid $\beta = \sigma_1^{-n_1} \cdots \sigma_k^{-n_k}$ closes up to give L^-.

Observe the easy identities: $\frac{t+1}{\sqrt{t}} = \frac{t^{-1}+1}{\sqrt{t^{-1}}}$, $\Phi_n(\sigma_i^{-1}) = (t^{-1} + 1)e_i - 1$ and $\frac{t}{(t+1)^2} = \frac{t^{-1}}{(t^{-1}+1)^2}$. It follows readily that the Jones polynomial of the mirror image link is $V_L(t^{-1})$. For example, the right-handed trefoil knot has polynomial $-t^{-4} + t^{-3} + t^{-1}$.

8   Relation to ring extension theory

Bruno Müller published in [13, 1964-5] his theory of quasi-Frobenius extensions, in which various phenomena observable in this paper first appeared. We establish in the next theorem that the basic construction of a finite separable extension is none other than the endomorphism ring of the natural module associated to the extension. Then the inclusion of A in A_1 is simply the left regular representation of A in End A_S. For a quasi-Frobenius extension $A \supseteq B$, it was shown in [13] that the endomorphism ring End A_B is itself a quasi-Frobenius extension of A: then the converse question was taken up and settled by Müller. We explore what finite separability has to do with quasi-Frobenius extension in this section.

Proposition 8.1 A_1 is Morita equivalent to S.

Proof. Recall that $e_1 = 1 \otimes_S 1$, an idempotent in A_1. Define a ring homomorphism,

$$F : \quad A_1 \to \text{End } A_S$$

by

$$F(\sum_{i=1}^{m} a_i e_1 b_i) = \sum_{i=1}^{m} \lambda_{a_i} E \lambda_{b_i}$$

where $\lambda_x(y) = xy$ is the left multiplication map. It is easy to show that F is surjective, and slightly harder to show that it is injective. But A_S is a generator module since $E(1) = 1$ and finitely generated projective by an earlier result. It follows from basic Morita theory that A_1 and S are equivalent rings. $\square$

A right quasi-Frobenius extension is a ring A and subring B containing 1_A subject to two axioms:

1. A_B is a finitely generated projective module;

2. A is isomorphic as $B - A$ modules to a direct summand of a direct sum of a finite number of copies of $\hom_B(A_B, B_B)$.

Theorem 8.1 A finite separable extension $A \supseteq S$ is a right quasi-Frobenius extension.

Proof. We have seen that the natural module A_S is a finitely generated projective.

Define an $S - A$ isomorphism $\Psi : A \to \hom({}_A A_S, S_S)$ by

$$a \longmapsto E(a-)$$

where $E(a-)$ denotes the map $x \mapsto E(ax)$. Since E is nondegenerate (corollary, section 2), Ψ is injective. Ψ is surjective since, given a map f in the range, $\Psi : \sum_{i=1}^{n} f(x_i)y_i \mapsto f$. The $S - A$ action on $\hom(A_S, S_S)$ is given by the formula $(sga)(x) = s[g(ax)]$. Now it is clear that $\Psi(sab) = s\Psi(a)b$ $\square$ [3]

It is not true that every quasi-Frobenius extension is a finite separable extension. Indeed the next theorem can only be proven for finite separable extensions. A simple example of a Frobenius algebra (i.e., the paradigm of a quasi-Frobenius extension) not satisfying equality of global dimension of rings, $D(-)$, is a group algebra of a p-group in characteristic p.

Theorem 8.2 If A is a finite separable extension of S, then

$$D(A) = D(S).$$

Proof. D(-) may denote left, right, or weak global dimension of ring, though we give the argument only for left modules. Let pd denotes projective dimension. Recall the well-known inequality for any change of rings $S \to A$ and an A-module M, pd $M_S \leq$ pd $M_A +$ pd A_S. Since the inclusion map of

---

[3]Left quasi-Frobenius extension may be defined oppositely. The inclusion in Theorem 8.1 is true for these as well.

$N \to N \otimes_S A$ is split by $Id \otimes_S E$ whenever A is a split extension of S, we easily see that for split extensions

$$D(S) \le D(A) + \text{pd } A_S.$$

For finite separable extension we see moreover that $D(S) \le D(A)$, since pd $A_S = 0$. Since A_1 is a finite separable extension of A we have $D(A) \le D(A_1)$. But A_1 is Morita equivalent to S so that $D(A_1) = D(S)$. □

The last theorem is a generalization of Serre's extension theorem for group cohomology [17, the coprime index case]. For we recall that the cohomological dimension of a group G over a field k

$$\text{cd}_k(G) = \text{D}(k[G]).$$

References

[1] M. Auslander and O. Goldman, The Brauer group of a commutative ring, *Trans. A.M.S.* **97** (1960), 367-409.

[2] V. L. Hansen, Braids and Coverings: Selected Topics, *London Math. Soc. Student Text* **18**, Cambridge Univ. Press, 1989.

[3] K. Hirata and K. Sugano, On semisimple extensions and separable extensions over noncommutative rings, *J. Math. Soc. Japan* **18** (1966), 360-373.

[4] V.F.R. Jones, Index for subrings of rings, Contemp. Math. **43**, AMS (1985), 181-190.

[5] V.F.R. Jones, Hecke algebra representations of braid groups and link polynomials, *Ann. of Math.,* **126** (1987), 335-388.

[6] V.F.R. Jones, Subfactors and Knots, CBMS Reg. Conf. Ser. No. **80**, AMS, 1991.

[7] G. Hochschild, On the cohomology groups of an associative algebra, *Annals of Math.* **46** (1945), 58-67.

[8] G. Hochschild, Relative homological algebra, *Trans. AMS* **82** (1956), 246-269.

[9] L. Kadison, Algebraic aspects of the Jones basic construction, *C.R. Math. Rep. Acad. Sci. Canada,* to appear.

[10] L. Kadison and D. Kastler, Cohomological aspects and relative separability of finite Jones index subfactors, *Nachr. Akad. Wissen. Göttingen,* II, nr. 4 (1992), 1-11.

[11] I. Kaplansky, Fields and Rings, 2nd Ed., Univ. of Chicago Press, 1972.

[12] S. Lang, Algebra, Addison-Wesley, 6th printing, Reading, Mass., 1974.

[13] B. Müller, Quasi-Frobenius Erweiterungen, *Math Z.* **85** (1964), 345-368. Quasi-Frobenius Erweiterungen II, *Math Z.* **88** (1965), 380-409.

[14] D.S. Passman, The algebraic structure of group rings, Wiley Interscience, New York, 1977.

[15] R. Pierce, Associative Algebras, Graduate Texts in Math. **88**, Springer Verl., 1982.

[16] M. Pimsner and S. Popa, Entropy and the index of subfactors, *Annales Scientifique Ecole Normale Superieur* (1986), 57-106.

[17] J.-P. Serre, Cohomologie des groupes discrets, *Prospects in Math., Annals of Math. Studies* **70** (1970), 77-169.

A Note on Gröbner Bases and Reduced Ideals

T. KAMBAYASHI

Department of Mathematical Sciences, Tokyo Denki University

Hatoyama, Saitama-ken 350–03, JAPAN

In this note we work with sets and ideals in a polynomial ring $K^{[n]} := K[X_1, ..., X_n]$ in n variables over a field K. So, in particular, by an "ideal" we always mean one contained in $K^{[n]}$. Our aim is to give a sufficient condition for an ideal to be reduced, *i.e.*, to be its own radical, in terms of its Gröbner basis (*cf.* Theorem below).

Let $\leq$ be a total order on the set S of all power-products $X_1^{i(1)} X_2^{i(2)} ... X_n^{i(n)}$ in the X's. Recall from the theory of Gröbner Bases (Buchberger, 1985; Gianni *et al.*, 1989; Abhyankar, 1989) that the order $\leq$ is defined to be *admissible* if (i) $1 < p$ holds for all $p \in S$, and (ii) $p \leq q$ implies $p \cdot r \leq q \cdot r$ for all $p, q, r \in S$. We shall fix an admissible total order $\leq$ on S once and for all. Also, for convenience's sake, we shall enlarge the order relation $\leq$ to the set of *all* monomials by saying $a \cdot p \leq b \cdot q$ whenever $p \leq q$ in S with $a, b \in K - \{0\}$.

For any nonzero polynomial $f \in K^{[n]}$ we define its *leading term*, denoted $\ell t(f)$, to be that term of f which is the largest among the terms of f with respect to the enlarged $\leq$ defined just above. We define $\ell t(0) = 0$. For any set $E \subset K^{[n]}$, we denote by $< E >$ the ideal generated by E, and by $Lt(E)$ the ideal $<\ell t(f) : f \in E>$ generated by all leading terms of the elements of E. Note that, in general, $Lt(E)$ is contained in $Lt(< E >)$ without being equal to it. [Consider, for example, the case

$E = \{ 1 + x^2, xy \}, 1 < x < y$. Then, $y \in \mathrm{Lt}(<E>), y \notin \mathrm{Lt}(E)$.]

Lemma 1. *Let I be an ideal generated by monomials. Then,*
(i) *any monomial belonging to I is divisible by one of the generating monomials; and*
(ii) *a polynomial f belongs to I if and only if every term of f belongs to it.*

This follows easily when one regards I as a vector space over K, which then is spanned by all monomials divisible by one of the generating monomials. $\square$

Let J be an ideal. A subset $G \subset J$ is called a *Gröbner Basis of J* if $\mathrm{Lt}(J) = \mathrm{Lt}(G)$. [This definition appears to originate in Möller–Mora (1986). It is different from, but equivalent to, the original one by Buchberger (1985).] One can see at once that, if G is a Gröbner basis for J, then $J = <G>$, or G generates J.

An ideal J is said to be *reduced* if J agrees with its own radical; or, equivalently, if $f^2 \in J$ entails $f \in J$ always.   Our first result is the following:

Proposition 2. *For any ideal $J \subset K^{[n]}$, if $\mathrm{Lt}(J)$ is reduced then J is reduced.*

Proof. Suppose J not reduced, and consider the polynomials h such that $h \notin J, h^2 \in J$. Among all such h's let f be one whose leading term $\ell t(f)$ is minimal.   Then, $(\ell t(f))^2 = \ell t(f^2) \in \mathrm{Lt}(J)$ is clear .   But, $\ell t(f) \notin \mathrm{Lt}(J)$. For, if $\mathrm{lt}(f) \in \mathrm{Lt}(J)$, then by Lemma 1 $\ell t(f) = \ell t(g)$ for some $g \in J$. But then $\ell t(f - g) < \ell t(f)$, whereas $f - g \notin J, (f - g)^2 \in J$. This is contrary to the minimality of f among the h's as above. $\square$

Simple examples such as $J = <x^2 + y>, 1 < y < x$ show that the converse of Prop. 2 does not hold.

Proposition 3. *Let J be an ideal in $K^{[n]}$ generated by monomials. If each of the generating monomials is square-free, then J is reduced.*

Proof. Let $J = <m_1, ..., m_q>$ with square-free monomials $m_i, 1 \leq i \leq q$.   Assume there is an $h \notin J, h^2 \in J$.   Among such polynomials h,   take f to be one with the lowest leading term relative to the $\leq$. Now, if we have $\ell t(f) \in J$, then $f - \ell t(f) \notin J$ and $(f - \ell t(f))^2 \in J$, and $f - \ell t(f)$ has a strictly lower leading term than f. This goes against our initial choice of f. So, $\ell t(f) \notin J$ must hold, to begin with. Now write $f = a \cdot p + g$ with $\ell t(f) = a \cdot p, a \in K - 0, p \in S$. Then, $\ell t(f^2) = a^2 \cdot p^2 = (\ell t(f))^2 \in J = <m_{1,...,}m_q>$ by Lemma 1.   It follows, also by Lemma 1, that p^2 is divisible by one of the m_i's, say m_j.   Since m_j is square-free, this implies p is divisible by m_j, so that $\ell t(f) = a \cdot p \in J$. A contradiction. $\square$

By combining Prop. 2 with Prop. 3 we now get the main result:

Theorem. *Let $J \subset K^{[n]}$ be an ideal, and E a Gröbner basis for J, with respect to a given admissible order $\leq$ on S. If for each $g \in E$ its leading term $\ell t(g)$ is square-free, then J is reduced.* $\Box$

Let B be a Gröbner basis for an ideal J. Suppose, for some $f \neq g$ both in B, $lt(f)$ is divisible by $lt(g)$. Clearly, then, the set $B - \{f\}$, too, is a Gröbner basis for J. Continuing to throw out redundant members like f from B , one will reach a *minimal* Gröbner basis, whose obvious definition we shall omit here. Observe that the theorem above gains its full force when applied to minimal Gröbner bases.

References

Abhyankar, S. S. (1989). On the Jacobian Conjecture: A new approach *via* Gröbner Bases, *J. Pure Applied Alg.*, **61**: 211

Buchberger, B. (1985). Gröbner bases: an algorithmic method in polynomial ideal theory, MULTIDIMENSIONAL SYSTEM THEORY (H. K. Bose, *ed.*), Reidel, Dordrecht, Chapter 6.

Gianni, P., Trager, B., and Zacharias, G. (1989). Gröbner bases and primary decomposition of polynomial ideals, COMPUTATIONAL ASPECTS OF COMM. ALG. (L. Robbiano, *ed.*), Academic Press, London and San Diego, p.15.

Möller, H. M. and Mora, F. (1986). New constructive methods in classical ideal theory, *J. Algebra*, **100**: 138.

Bicomplexes and Galois Cohomology

H. F. KREIMER   Department of Mathematics,
Florida State University, Tallahassee, Florida

In this paper, a new derivation of the well known long exact sequence of Galois cohomology for commutative rings is presented. An attempt to incorporate much of what is known about the terms and mappings of that exact sequence into a single theory is made. A filtration of a differential, graded module gives rise not only to an exact couple and a spectral sequence, but to a more elaborate structure herein called an exact octahedron. Long exact sequences are obtained by unwinding "strands" of an exact octahedron, and relationships between the exact sequences are recorded in the exact octahedron.

MacLane (1963), especially section 5 of chapter XI on exact couples, is a helpful resource; and a very readable account of the technique of faithfully flat descent, which will be used to interpret cohomology groups, is found in Knus and Ojanguren (1974). Unless otherwise specified by notation or context, all modules, and homomorphisms of modules are over a given commutative ring R.

1. BICOMPLEXES

A bicomplex X is a bigraded module with homogeneous homomorphisms ∂' of bidegree $(-1, 0)$ and ∂'' of bidegree $(0, -1)$ such that

$$\partial'\partial' = 0, \quad \partial'\partial'' + \partial''\partial' = 0, \quad \partial''\partial'' = 0. \tag{1.1}$$

Thus a bicomplex consists of a family $\{X_{p,q}\}$ of modules and two families, $\partial' : X_{p,q} \to X_{p-1,q}$ and $\partial'' : X_{p,q} \to X_{p,q-1}$, of module homomorphisms, defined for all integers p and q and for which conditions (1.1) are satisfied. The totalization of X is a complex in which $(\text{Tot } X)_n$ is the direct sum $\sum_{p+q=n} X_{p,q}$ and the differential operator is $\partial' + \partial''$. The first filtration of subcomplexes of Tot X is defined by setting $(F_p \text{Tot} X)_n = \sum_{r \leq p} X_{r,n-r}$. It is noteworthy that the family of mappings $(-1)^q \partial' : X_{p,q} \to X_{p-1,q}$ yields a chain transformation between the quotient complexes $F_p \text{Tot } X/F_{p-1} \text{Tot } X$ and $F_{p-1} \text{Tot } X/F_{p-2} \text{Tot } X$, and $F_p \text{Tot } X/F_{p-2} \text{Tot } X$ is the mapping cone of this chain transformation.

In a more general setting, let F be a filtration of a differential, graded module A; and denote the homology groups of subcomplexes and quotient complexes by $D_{p,q} = H_{p+q}(F_p A)$, $E_{p,q} = H_{p+q}(F_p A/F_{p-1}A)$, and $K_{p,q} = H_{p+q}(F_p A/F_{p-2}A)$, for integers p and q. Three exact couples, $(D, E; i, j, k)$, $(E, K; jk, e, f)$, and $(D, K; i^2, g, h)$, in which the homomorphisms h and k are induced by the differential operator on A and the homomorphisms e, f, g, i, and j are induced by the identity map on A, are obtained as long exact sequences of homology from the short exact sequences of complexes: $F_{p-1}A \to F_p A \to F_p A/F_{p-1}A$, $\quad F_{p-1}A/F_{p-2}A \to F_p A/F_{p-2}A \to F_p A/F_{p-1}A$, and $F_{p-2}A \to F_p A \to F_p A/F_{p-2}A$, respectively. Moreover, the following diagram of bigraded modules is commutative.

$$
\begin{array}{ccccc}
E & \xrightarrow{e} & K & \xrightarrow{f} & E \\
k \downarrow & \diagdown^{h} & & & k \downarrow \\
D & & \xrightarrow{i} & & D \\
j \downarrow & & & {}^{g}\diagup\, j \downarrow \\
E & \xrightarrow{e} & K & \xrightarrow{f} & E
\end{array}
$$

These data may be summarized by the following octahedron of bigraded modules, in which four of the eight faces are exact couples (the exact couple $(D, E; i, j, k)$ appears twice), and the remaining four faces and two of the three interior squares are commutative diagrams

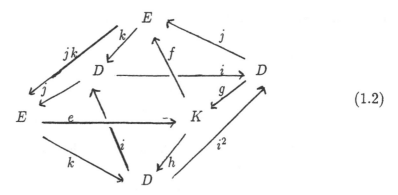

$$(1.2)$$

Such an octahedron will be called exact.

From an exact couple $(D, E; i, j, k)$, a derived exact couple $(D', E'; i', j', k')$ is obtained by setting $D' = \operatorname{Im} i$ and $E' = \operatorname{Ker} jk/\operatorname{Im} jk$, letting the mappings i' and k' be induced by the mappings i and k, and letting the mapping j' be induced by the additive relation ji^{-1}. Crucial to obtaining a derived exact octahedron is a construction of Villamayor and Zelinsky (1977) for the following commutative diagram.

$$(1.3)$$

LEMMA 1.4. (Villamayor-Zelinsky). *Let there be given a commutative diagram of graded modules as in figure 1.3, for which $\eta\gamma = 0$ and the sequence $E \xrightarrow{d} E \xrightarrow{e} K \xrightarrow{f} \overline{E} \xrightarrow{\overline{d}} \overline{E}$, is exact. Then $dd = 0 = \overline{d}\,\overline{d}$; and for $E' = \operatorname{Ker} d/\operatorname{Im} d$, $\overline{E}' = \operatorname{Ker} \overline{d}/\operatorname{Im} \overline{d}$, and $K' = \operatorname{Ker} \eta/\operatorname{Im} \gamma$, there is a long exact sequence: $\to \overline{E}' \xrightarrow{d'} E' \xrightarrow{e'} K' \xrightarrow{f'} \overline{E}' \xrightarrow{d'} E' \to$, in which the homomorphisms d', e', and f' are induced by the additive relation ηf^{-1} and the homomorphisms e and f, respectively.*

Proof. Since $d\eta = \eta e\eta = \eta\gamma f = 0$, $dd = d\eta e = 0$. Likewise $\gamma\overline{d} = \gamma f\gamma = e\eta\gamma = 0$, and therefore $\overline{d}\,\overline{d} = f\gamma\overline{d} = 0$. From the hypothesis of exactness follows the exactness of the rows in the following commutative diagram.

$$0 \;\rightarrow\; \text{Coker } d \;\overset{e}{\rightarrow}\; K \;\overset{f}{\rightarrow}\; \overline{E} \;\rightarrow\; \text{Coim } \overline{d} \;\rightarrow\; 0$$

$$d \downarrow \qquad\qquad \eta \downarrow \qquad \gamma \downarrow \qquad\qquad \overline{d} \downarrow$$

$$0 \;\rightarrow\; \text{Im } d \;\rightarrow\; E \;\overset{e}{\rightarrow}\; K \;\overset{f}{\rightarrow}\; \text{Ker } \overline{d} \;\rightarrow\; 0$$

By a diagram chase or a version of the strong four lemma, there are the following exact sequence of kernels: $0 \rightarrow E' \overset{e}{\rightarrow} \text{Ker } \eta \overset{f}{\rightarrow} \text{Ker } \gamma \rightarrow 0$, and the following exact sequence of cokernels: $0 \rightarrow \text{Coker } \eta \overset{e}{\rightarrow} \text{Coker } \gamma \overset{f}{\rightarrow} \overline{E}' \rightarrow 0$. Now consider the commutative diagram:

$$\text{Coker } \eta \;\overset{e}{\rightarrow}\; \text{Coker } \gamma \;\overset{f}{\rightarrow}\; \overline{E}'$$

$$d \downarrow \qquad\qquad \eta \downarrow \qquad\qquad \downarrow$$

$$\text{Ker } d \;\;=\;\; \text{Ker } d \;\rightarrow\; 0$$

The kernel-cokernel exact sequence for this diagram is: $0 \rightarrow \text{Ker } d / \text{Im } \eta \rightarrow K' \rightarrow \overline{E}' \rightarrow E' \rightarrow \text{Ker } d / \text{Im } \eta$, and the Yoneda composite of these exact sequences is the desired long exact sequence.

Note that the additive relations ηf^{-1} and $e^{-1}\gamma$ are the same, and the lemma could be proved also by considering the kernel-cokernel exact sequence for the following commutative diagram.

$$0 \;\rightarrow\; \text{Coker } \overline{d} \;\;=\;\; \text{Coker } \overline{d}$$

$$\downarrow \qquad\qquad \gamma \downarrow \qquad\qquad \overline{d} \downarrow$$

$$E' \;\overset{e}{\rightarrow}\; \text{Ker } \eta \;\overset{f}{\rightarrow}\; \text{Ker } \gamma$$

$$\square$$

PROPOSITION 1.5. *If D, E, and K are terms and e, f, g, h, i, j, and k are mappings of an exact octahedron as in figure 1.2; then another exact octahedron is derived by setting $D' = \text{Im } i$, $E' = \text{Ker } jk / \text{Im } jk$ and $K' = \text{Ker } jh / \text{Im } gk$, letting the mappings e', f', h', i', and k', be induced by the mappings e, f, h, i, and k, and letting the mappings g' and j' be induced by the additive relations gi^{-1} and ji^{-1}.*

Proof. The exactness of the couple $(E', K', j'k', e', f')$ follows from Lemma 1.4 by setting $\overline{E} = E$, $\overline{d} = d = jk$, $\gamma = gk$, and $\eta = jh$, and noting that $j'k'$

is the homomorphism of E' into E' induced by the additive relation $\eta f^{-1}(=$ $jhf^{-1} = ji^{-1}k = e^{-1}gk)$. $(D', K'; (i')^2, g', h')$ is shown to be an exact couple by noting that the domain of definition of the additive relation gi^{-1} is Im $i =$ D', the indeterminancy of gi^{-1} is $g(\text{Ker } i) = g(\text{Im } k) = \text{Im } gk$, Ker $gi^{-1} =$ $i(\text{Ker } g) = i(\text{Im } i^2) = i^2(D')$, and Im $gi^{-1} = \text{Im } g = \text{Ker } h$. Also $h(\text{Ker } jh) =$ $hh^{-1}(\text{Ker } j) = \text{Ker } j \cap \text{Im } h = \text{Im } i \cap \text{Ker } i^2 = D' \cap \text{Ker } i^2$. The required conditions of commutativity are readily verified. $\qquad\square$

2. Amitsur Cohomology

Let R be a commutative ring, let S be a commutative algebra over R, let $S^0 = R$, and let $S^{n+1} = S^n \otimes_R S$ for each nonnegative integer n. There are $n + 1$ homomorphisms of the algebra S^n into the algebra S^{n+1}; and they may be defined recursively as follows: let ε_0 be the canonical homomorphism of R into the R-algebra S; and for each positive integer n, use again the symbol ε_k to denote the natural extension of the homomorphism ε_k of S^{n-1} into S^n to a homomorphism $\varepsilon_k \otimes S$ of $S^n = S^{n-1} \otimes_R S$ into $S^{n+1} = S^n \otimes_R S$, $0 \le k \le n-1$, and let ε_n be the homomorphism of S^n onto the first factor of $S^n \otimes_R S = S^{n+1}$. The homomorphism ε_k of S^n into S^{n+1} determines a functor from the category of S^n-modules to the category of S^{n+1}-modules, and this functor will also be denoted by the symbol ε_k. Note that $\varepsilon_i \varepsilon_j = \varepsilon_{j+1} \varepsilon_i$ for $i \le j$.

If F is a functor from the category of commutative algebras over R to the category of abelian groups, the Amitsur complex $C(S/R, F)$ is a cocomplex with terms $F(S^n)$ for $n \ge 1$. The coboundary from $F(S^n)$ into $F(S^{n+1})$ is the alternating sum $\sum_{k=0}^{n}(-1)^k F(\varepsilon_k)$: and the Amitsur cohomology groups $H^n(S/R, F)$, $n \ge 0$, are the cohomology groups of this complex. Let U denote the functor which assigns to a ring its multiplicative group of invertible elements, and let P denote the functor which assigns to a commutative ring its Picard group of isomorphism classes of rank one, projective modules.

If T is a commutative algebra over R, there is a third quadrant bicomplex $X = C(S, T/R, U)$ with terms $X^{p,q} = U(S^{p+1} \otimes_R T^{q+1})$, for $p \ge 0$ and $q \ge 0$, and $X^{p,q} = 0$, otherwise. The convention of using positive superscripts instead of negative subscripts is being followed here. The homomorphism $(-1)^q \partial'$ of $U(S^{p+1} \otimes_R T^{q+1})$ into $U(S^{p+2} \otimes_R T^{q+1})$ is the coboundary of the Amitsur complex $C(S \otimes_R T^{q+1}/T^{q+1}, U)$, and the homomorphism ∂'' of $U(S^{p+1} \otimes_R T^{q+1})$ into $U(S^{p+1} \otimes_R T^{q+2})$ is the coboundary of the Amitsur complex $C(S^{p+1} \otimes_R T/S^{p+1}, U)$. Then with respect to the first filtration of the bicomplex, $F^p\text{Tot } X/F^{p+1}\text{Tot}$ is the Amitsur complex $C(S^{p+1} \otimes_R T/S^{p+1}, U)$; and for the associated exact octahedron, $E^{p,q} = H^q(S^{p+1} \otimes_R T/S^{p+1}, U)$.

From now on, assume that T is a faithfully flat R-module. Then $S^{p+1} \otimes_R T$ is a faithfully flat module over S^{p+1}, and $E^{p,0} = H^0(S^{p+1} \otimes_R T/S^{p+1}, U) = U(S^{p+1})$ (Knus and Ojanguren, 1974, Chapter V, Proposition 2.1). To the homomorphism of an R-algebra S onto the first factor of $S \otimes_R T$, there corresponds a group homomorphism of $P(S)$ into $P(S \otimes_R T)$. Letting $KP(S, T)$ denote the kernel of this group homomorphism, $KP(\cdot, T)$ is a functor from the category of R-algebras to the category of abelian groups and $E^{p,1} = H^1(S^{p+1} \otimes_R T/S^{p+1}, U) = KP(S^{p+1}, T)$ by (Knus and Ojanguren, 1974, Chapter V, Proposition 2.1). To give a brief sketch of the argument for this last assertion, regard $S^{p+1} \otimes_R T^2$ as a module over the algebra $S^{p+1} \otimes_R T^2$ and an element u of $U(S^{p+1} \otimes_R T^2)$ as an automorphism of this module. The condition that u be a cocycle of the Amitsur complex $C(S^{p+1} \otimes_R T/S^{p+1}, U)$ completes the descent data necessary to obtain an S^{p+1}-module M such that $M \otimes_R T$ and $S^{p+1} \otimes_R T$ are isomorphic $S^{p+1} \otimes_R T$-modules; and M must be a rank one, projective S^{p+1}-module, since T is faithfully flat over R. If M and M' are S^{p+1}-modules such that $M \otimes_R T$ and $M' \otimes_R T$ are isomorphic to $S^{p+1} \otimes_R T$, then an isomorphism of M onto M' may be lifted to an automorphism of $S^{p+1} \otimes_R T$ and an element of $U(S^{p+1} \otimes_R T)$. It follows readily that cohomologous cocycles determine isomorphic S^{p+1}-modules.

For the associated exact octahedron, the homomorphism jk of $E^{p,q}$ into $E^{p+1,q}$ is induced by $(-1)^q \partial'$; and for the derived exact octahedron, the terms and mappings of which will be designated with the subscript 2, $E_2^{p,0} = H^p(S/R, U)$ and $E_2^{p,1} = H^p(S/R, KP(\cdot, T))$. From the derived exact octahedron, the following long exact sequences are obtained.

$$E_2^{p-1,q} \to E_2^{p+1,q-1} \to K_2^{p,q} \to E_2^{p,q} \to E_2^{p+2,q-1} \to K_2^{p+1,q} \to \qquad (2.1)$$

Set $p = 0$ and $q = 1$ to obtain the long exact sequence:

$$\begin{aligned} 0 \to H^1(S/R, U) \to K_2^{0,1} &\to H^0(S/R, KP(\cdot, T)) \to H^2(S/R, U) \\ &\to K_2^{1,1} \to H^1(S/R, KP(\cdot, T)) \to H^3(S/R, U) \to \dots \end{aligned} \qquad (2.2)$$

It remains to identify $K_2^{0,1}$ and $K_2^{1,1}$.

A 1-cocycle in the cocomplex $F^p X/F^{p+2} X$ is an ordered pair (u, v) in $U(S^{p+1} \otimes_R T^2) \oplus U(S^{p+2} \otimes_R T)$ such that $\partial'' u = 0$ and $\partial' u = \partial'' v$. These mappings are displayed in the following diagram.

$$U(S^{p+2} \otimes_R T) \xleftarrow{\partial'} U(S^{p+1} \otimes_R T)$$
$$\partial'' \downarrow \qquad\qquad \partial'' \downarrow$$
$$\xleftarrow{\partial'} U(S^{p+1} \otimes_R T^2)$$
$$\partial'' \downarrow$$

Just as before, the condition $\partial'' u = 0$ completes descent data necessary to obtain an S^{p+1}-module M such that $M \otimes_R T$ and $S^{p+1} \otimes_R T$ are isomorphic $S^{p+1} \otimes_R T$-modules. When $p = 0$, the element v of $U(S^2 \otimes_R T)$ is the "lifting" of an isomorphism of $\varepsilon_0(M)$ onto $\varepsilon_1(M)$, where $\varepsilon_0(M)$ and $\varepsilon_1(M)$ are the S^2-modules $S \otimes_R M$ and $M \otimes_R S$, respectively. Thus to the 1-cocycle (u, v) there corresponds a system (M, f) consisting of a rank one, projective S-module M which is split by $S \otimes_R T$ and an isomorphism f of the S^2-module $S \otimes_R M$ onto the S^2-module $M \otimes_R S$. A 1-coboundary has the form $(\partial'' w, \partial' w)$ for an element w of $U(S \otimes_R T)$, and w may be regarded as the "lifting" of an isomorphism φ which is defined for the S-module M and carries the system (M, f) to a system $(\varphi(M), \varepsilon_1(\varphi) \circ f \circ \varepsilon_0(\varphi)^{-1})$. Therefore the elements of $K^{0,1}$ are isomorphism classes of such systems. The homomorphism jh of $K^{0,1}$ into $E^{2,0} = U(S^3)$ is induced by ∂', and the isomorphism class of (M, f) is in the kernel of jh when $\varepsilon_2(f) \circ \varepsilon_0(f) = \varepsilon_1(f)$. But the S-module M and the S^2-module isomorphism f of $S \otimes_R M$ onto $M \otimes_R S$ such that $\varepsilon_2(f) \circ \varepsilon_0(f) = \varepsilon_1(f)$ are descent data. If S is a faithfully flat module over R, then $M = N \otimes_R S$ for some R-module N and $N \otimes_R S \otimes_R T = M \otimes_R T$ is isomorphic to $S \otimes_R T$. This completes a sketch of the essential steps in the proof of the following proposition.

PROPOSITION 2.3. *If the algebra S over R is a faithfully flat module over R, then $K_2^{0,1} = KP(R, S \otimes_R T)$.*

Now Let $p = 1$, and proceed as before *mutatis mutandis*. In this case, M is an S^2-module, which will be regarded as a left S, right S bimodule; f is an isomorphism of $\varepsilon_1(M)$ onto the tensor product $\varepsilon_2(M) \otimes \varepsilon_0(M)$ of the S^3-modules $\varepsilon_2(M)$ and $\varepsilon_0(M)$; and the isomorphism class of the system (M, f) is in the kernel of jh when $\varepsilon_0(f) \circ \varepsilon_2(f) = \varepsilon_3(f) \circ \varepsilon_1(f)$. In particular, $\varepsilon_0(M) = S \otimes_R M$, $\varepsilon_2(M) = M \otimes_R S$, and the tensor product $\varepsilon_2(M) \otimes \varepsilon_0(M)$ is naturally isomorphic to and may be identified with $M \otimes_S M$. A comultiplication, which is a bimodule homomorphism of M into $M \otimes_S M$, is obtained by composing the canonical mapping of M into $\varepsilon_1(M)$ with these isomorphisms. It is convenient to denote the image of an element m of M under this comultiplication by $\sum_{(m)} m_{(1)} \otimes m_{(2)}$. The S^3-module $\varepsilon_1(M)$ may be identified with $M \otimes_R S$, where the first and third factors of S^3 act on M and the second factor of S^3 acts on S; and the isomorphism of $\varepsilon_1(M)$ onto $M \otimes_S M$ maps an element $m \otimes s$ of $M \otimes_R S$ onto $\sum_{(m)} m_{(1)}s \otimes m_{(2)}$. The equation $\varepsilon_0(f) \circ \varepsilon_2(f) = \varepsilon_3(f) \circ \varepsilon_1(f)$ is the condition of commutativity for the following diagram of S^4-modules, with tensor products taken over S^4 unless otherwise specified.

$$
\begin{array}{ccc}
\varepsilon_2\varepsilon_1(M) = \varepsilon_1\varepsilon_1(M) & \overset{\varepsilon_1(f)}{\to} & \varepsilon_1\varepsilon_2(M) \otimes \varepsilon_1\varepsilon_0(M) = \varepsilon_3\varepsilon_1(M)\otimes\varepsilon_0\varepsilon_0(M) \\
\varepsilon_2(f)\downarrow & & \| \\
\varepsilon_2\varepsilon_2(M)\otimes\varepsilon_2\varepsilon_0(M) & & \varepsilon_1(M)\otimes_S M \\
\| & & \varepsilon_3(f)\downarrow \\
\varepsilon_3\varepsilon_2(M)\otimes\varepsilon_0\varepsilon_1(M) = M \otimes_S \varepsilon_1(M) & \overset{\varepsilon_0(f)}{\to} & M\otimes_S M \otimes_S M
\end{array}
$$

It is readily apparent that the commutativity of the preceding diagram is equivalent to the coassociativity of the comultiplication. Finally, the homomorphism gk of $E^{0,1} = KP(S,T)$ into $K^{1,1}$ is induced by ∂', and it may be described in terms of representatives of isomorphism classes as follows. Let N be a rank one, projective S-module and let $N^* = \mathrm{Hom}_S(N, S)$. Then $N \otimes_S N^*$ is isomorphic to $\mathrm{Hom}_S(N, N)$, which is S since N is a rank one, projective S-module. The tensor product of the S^2-modules $N^* \otimes_R N$ and $N \otimes_R N^*$ is naturally isomorphic to $(N \otimes_S N^*) \otimes_R (N \otimes_S N^*) = S \otimes_R S$; and therefore, $N^* \otimes_R N$ is a rank one, projective S^2-module. Moreover $(N^* \otimes_R N) \otimes_S (N^* \otimes_R N)$ is isomorphic to $N^* \otimes_R S \otimes_R N$, which is $\varepsilon_1(N^* \otimes_R N)$; and thus an element of $K^{1,1}$ is determined.

Thus far the description of $K^{1,1}$ does not look familiar; but let $A = \mathrm{Hom}_{-S}(M, S)$ be the dual of the right S-module M for the system (M, f). There is a sequence of natural homomorphisms:

$$A \otimes_S A = A \otimes_S \mathrm{Hom}_{-S}(MS) \to \mathrm{Hom}_{-S}(M, A)$$
$$= \mathrm{Hom}_{-S}(M, \mathrm{Hom}_{-S}(M, S))$$
$$\approx \mathrm{Hom}_{-S}(M \otimes_S M, S),$$

under which the mapping $x \otimes y \to \varphi(\psi(x) \cdot y)$ of $M \otimes_S M$ into S corresponds to an element $\varphi \otimes \psi$ of $A \otimes_S A$. The comultiplication of M determines a multiplica-

tion on A by the rule: $\varphi \times \psi(m) = \sum_{(m)} \varphi(\psi(m_{(1)}) \cdot m_{(2)})$. There is a natural homomorphism, which is an isomorphism when S is a finitely generated, projec-
tive module over R, of $A \otimes_R S = \text{Hom}_{-S}(M, S) \otimes_R S$ into $\text{Hom}_{-S}(M, S \otimes_R S)$,
where the right S-module structure of $S \otimes_R S$ arises from the first factor
S. Since M is a rank one, projective S^2-module, $\text{Hom}_{S^2}(M, M) = S^2$; and
$\text{Hom}_{-S}(M, S \otimes_R S) = \text{Hom}_{-S}(M, \text{Hom}_{S^2}(M, M))$, which is isomorphic to
$\text{Hom}_{-S^2}(M \otimes_S M, M)$ by adjoint associativity. The isomorphism f induces an
isomorphism of $\text{Hom}_{-S^2}(M \otimes_S M, M)$ onto $\text{Hom}_{-S^2}(\varepsilon_1(M), M)$, which is naturally isomorphic to $\text{Hom}_{-S}(M, M)$. Through this sequence of mappings, the el-
ement $\varphi \otimes s$ of $A \otimes_R S$ acts on M by the rule: $(\varphi \otimes s) \cdot m = \sum_{(m)} \varphi(m_{(1)}) \cdot m_{(2)} \cdot s$;
and the mapping of $A \otimes_R S$ into $\text{Hom}_{-S}(M, M)$ is a homomorphism of asso-
ciative algebras when the comultiplication on M is coassociative. Now assume
that S is a faithful, finitely generated, and projective module over R, and the
comultiplication on M is coassociative. Then M is a faithful, finitely generated, projective right S-module; $\text{Hom}_{-S}(M, M)$ is a central, separable algebra
over S; and A must be an algebra which has an identity element and is central
and separable over R, since S is a faithfully flat module over R. If N is a rank
one, projective module over S, note that $\text{Hom}_{-S}(N^* \otimes_R N, S)$ is isomorphic
to $\text{Hom}_R(N^*, \text{Hom}_S(N, S)) = \text{Hom}_R(N^*, N^*)$ by adjoint associativity.

Conversely, suppose A is a central, separable algebra over R and V is
a faithful, finitely generated, projective S-module such that $A \otimes_R S$ and
$\text{Hom}_S(V, V)$ are isomorphic algebras over S. Letting $\Omega = \text{Hom}_S(V, V)$ and
$V^* = \text{Hom}_S(V, S)$, $V \otimes_S V^*$ is canonically isomorphic to Ω and the evaluation
mapping is an isomorphism of $V^* \otimes_\Omega V$ onto S. Identifying $A \otimes_R S$ with Ω, V
becomes a left A-module, V^* becomes a right A-module, and the S^2-module
$V^* \otimes_A V$ may be formed. The argument that $V^* \otimes_A V$ is a rank one, projective
module over S^2 is delicate. The S^2-module $V \otimes_R S$ becomes a module over
$\Omega = A \otimes_R S$ by letting A act on the factor V and S act on the factor S. Then
$V^* \otimes_\Omega (V \otimes S)$ is an S^2-module by the action of S^2 on the factor $V \otimes_R S$, and
$V \otimes_S V^* \otimes_\Omega (V \otimes_R S)$ is isomorphic to $\Omega \otimes_\Omega (V \otimes_R S) = V \otimes_R S$. Therefore
$(V^* \otimes_A V) \otimes_{S^2} (V^* \otimes_\Omega (V \otimes_R S))$ is isomorphic to $V^* \otimes_\Omega (V \otimes_R S)$, where
this time the actions of the factors A and S of $\Omega = A \otimes_R S$ are both on V.
But $V^* \otimes_\Omega V \otimes_R S = S \otimes_R S$, and this completes the argument that $V^* \otimes_A V$
is a rank one, projective S^2-module. Also note the sequence of isomorphisms:
$(V^* \otimes_A V) \otimes_S (V^* \otimes_A V) \approx V^* \otimes_A \Omega \otimes_A V = V^* \otimes_A (A \otimes_R S) \otimes_A V \approx$
$(V^* \otimes_A V) \otimes_R S = \varepsilon_1(V^* \otimes_A V)$. Moreover, $\text{Hom}_{-S}(V^* \otimes_A V, S)$ is isomorphic

to $\text{Hom}_A(V^*, V^*)$ by adjoint associativity; and it can be verified that this is an isomorphism of algebras over R. The right A-module V^* may be regarded as a left module over the opposite algebra A^{op}, and $\text{Hom}_A(V^*, V^*)$ is the centralizer of A^{op} in $\text{Hom}_R(V^*, V^*)$. Thus, if S is a faithful, finitely generated, projective module over R, $\text{Hom}_A(V^*, V^*)$ and A are equivalent central, separable R-algebras, (Auslander and Goldman, 1960b, Theorem 3.3 and Corollary 5.4). If $A = \text{Hom}_R(W, W)$ for a faithful, finitely generated, projective R-module W; then $V = W \otimes_R V_0$, where $V_0 = \overline{\text{Hom}}_R(W, R) \otimes_A V$ is a finitely generated, projective S-module which must have rank one over S, (Auslander and Goldman, 1960a, Appendix, Proposition A.4). Letting $V_0^* = \text{Hom}_S(V_0, S)$, $V^* = \text{Hom}_S(W \otimes_R V_0, S)$ is isomorphic to $\text{Hom}_R(W, V_0^*)$ by adjoint associativity, and the latter is isomorphic to $V_0^* \otimes_R \text{Hom}_R(W, R)$. Therefore $V^* \otimes_A V \approx V_0^* \otimes_R \text{Hom}_R(W, R) \otimes_A W \otimes_R V_0 \approx V_0^* \otimes_R R \otimes_R V_0 \approx V_0^* \otimes_R V_0$. The following proposition now can be asserted.

PROPOSITION 2.4. *If the algebra S over R is a faithful, finitely generated, projective module over R, then $K_2^{1,1}$ is isomorphic to a subgroup of the relative Brauer group $B(S/R)$ of equivalence classes of central, separable R-algebras which are split by S.*

The subgroup is determined by those central, separable R-algebras A which are split by S and for which $V^* \otimes_A V \otimes_R T$ and $S^2 \otimes_R T$ are isomorphic modules over $S^2 \otimes_R T$. This condition is awkward; but for any rank one, projective S^2-module M, there exists an R-algebra T which is a faithfully flat R-module and for which $M \otimes_R T$ and $S^2 \otimes_R T$ are isomorphic modules over $S^2 \otimes_R T$, (Chase and Rosenberg, 1965, Theorem 5.2). Thus $B(S/R)$ is the direct limit of $K_2^{1,1}$ over a suitable category of commutative R-algebras T which are faithfully flat modules over R (Chase and Rosenberg, 1965, Theorem 5.6). Also the functor P is a direct limit of the functors $KP(\cdot, T)$ over a suitable category of commutative R-algebras T. By taking direct limits, the following long exact sequence is obtained from 2.2 when S is a faithful, finitely generated, projective module over R.

$$0 \to H^1(S/R, U) \to P(R) \to H^0(S/R, P) \to H^2(S/R, U)$$
$$\to B(S/R) \to H^1(S/R, P) \to H^3(S/R, U) \to \overline{K}_2^{2,1} \to H^2(S/R, P) \to \ldots \tag{2.5}$$

This derivation of the exact sequence in 2.5 is merely a refinement of the methods used by Chase and Rosenberg (1965) to obtain the initial seven term exact sequence. Since exact sequences are preserved under direct limits, the exact sequence in 2.5 may be derived also by taking direct limits of the terms of the first exact octahedron to obtain an exact octahedron with terms $\overline{D}$, $\overline{E}$, $\overline{K}$, $\overline{E}^{p,0} = U(S^{p+1})$ and $\overline{E}^{p,1} = P(S^{p+1})$, and then forming the derived exact octahedron. This latter process is a variation of the techniques employed by Villamayor and Zelinsky (1977) to obtain the infinite exact sequence in 2.5. For each positive integer n, the groups J_n and $H^n(J)$, introduced in Villamayor and Zelinsky (1977), are isomorphic to the groups $\overline{K}^{n-1,1}$ and $\overline{K}_2^{n-1,1}$, respectively. Let $H^q(S,U) = \varinjlim_T H^q(S \otimes_R T/T, U)$ for a category of commutative R-algebras T which are faithfully flat modules over R. This definition of $H^q(S,U)$ does depend on the choice of the category.

THEOREM 2.6. *There is an exact octahedron with terms $\overline{D}$, $\overline{E}$, and $\overline{K}$, such that $\overline{E}^{p,q} = H^q(S^{p+1}, U)$ for nonnegative integers p and q ($\overline{E}^{p,q} = 0$ otherwise).*

It has been shown that $H^0(\cdot, U)$ and U are equivalent functors and $H^1(\cdot, U)$ and P are equivalent functors for suitably chosen categories of commutative R-algebras T. The canonical homomorphism of R into the R-algebra S determines a group homomorphism of $H^2(R,U)$ into $H^2(S,U)$, and it can be shown that $\overline{K}_2^{1,1}$ is the kernel of this group homomorphism, (Villamayor and Zelinsky, 1977, Theorem 6.14). Thus $H^2(R,U)$ is a generalization of the Brauer group of R. But the derived exact octahedron contains much more than one long exact sequence. In particular, there is the following commutative diagram with exact rows, by which the Amitsur cohomology groups with respect to the various functors $H^q(\cdot, U)$ are related.

$$
\begin{array}{ccccccc}
 & \downarrow & & \downarrow & & \| & \\
\rightarrow & H^{p-1}(S/R, H^{q+1}(\cdot, U)) & \rightarrow & \overline{K}_2^{p-2,q+2} & \rightarrow & H^{p-2}(S/R, H^{q+2}(\cdot, U)) & \rightarrow \\
 & \| & & \downarrow & & \downarrow & \\
\rightarrow & H^{p-1}(S/R, H^{q+1}(\cdot, U)) & \rightarrow & H^{p+1}(S/R, H^q(\cdot, U)) & \rightarrow & \overline{K}_2^{p,q+1} & \rightarrow \quad (2.7) \\
 & \downarrow & & \| & & \downarrow & \\
\rightarrow & \overline{K}_2^{p+1,q} & \rightarrow & H^{p+1}(S/R, H^q(\cdot, U)) & \rightarrow & H^{p+3}(S/R, H^{q-1}(\cdot, U)) & \rightarrow \\
 & \downarrow & & \downarrow & & \| & \\
\end{array}
$$

References

Auslander, M. and Goldman, O. (1960a). Maximal orders, *Trans. Amer. Math. Soc.*, **97**: 1-24.

Auslander, M. and Goldman, O. (1960b). The Brauer group of a commutative ring, *Trans. Amer. Math. Soc.*, **97**: 367-409.

Chase, S.U. and Rosenberg, A. (1965). Amitsur Cohomology and the Brauer group, *Memoirs A.M.S.*, **52**: 34-79.

Knus, M.A. and Ojanguren, M. (1974). Théorie de la descente et algèbres d'Azumaya, *Springer Lecture Notes*, **389**.

MacLane, S. (1963). *Homology*, Springer Verlag, Berlin.

Villamayor, O.E. and Zelinsky, D. (1977). Brauer groups and Amitsur cohomology for general commutative ring extensions, *Journal of Pure and Applied Algebra*, **10**: 19-55.

Adjoining Idempotents

ANDY R. MAGID   University of Oklahoma, Norman, Oklahoma

INTRODUCTION

Let R denote a commutative ring. Idempotents of R are elements $e \in R$ such that $e^2 = e$; examples are the additive and multiplicative identitiy elements 0 and 1. The set $B(R)$ of idempotents of R forms a Boolean algebra with operations:

$$e_1 \cap e_2 = e_1 e_2;$$
$$e_1 \cup e_2 = e_1 + e_2 - e_1 e_2;$$
$$e' = 1 - e;$$
$$\emptyset = 0;$$
$$\mathcal{U} = 1.$$

Now suppose $B \supseteq B(R)$ is a Boolean algebra extension. Is it the case that there is a ring R_B containing R such that $B(R_B) = B$ and so that the Boolean algebra inclusion $B(R_B) \supseteq B(R)$ coming from the ring inclusion $R_B \supseteq R$ is the given extension $B \supseteq B(R)$? More precisely, is there a commutative R algebra $R \to R_B$ and a Boolean algebra ismorphism $B(R_B) \to B$ such that the following commutes?

$$B(R_B) \longrightarrow B$$

$$B(R)$$

We will show in this paper that the above question has a positive answer, and furthermore discuss why the question is of interest. Our notation for R_B, which seems to depend only on R and B, is not yet accurate: this will be remedied by making it universal with respect to the above property. Then we will also see that R_B has a number of nice features, including being generated over R by idempotents and being a directed union of subalgebras isomorphic to products of copies of direct factors of R.

THE BOOLEAN SPECTRUM

In this section, we will review the Boolean spectrum, or space of connected components, of a commutative ring. This notion is due to Pierce [P], with additional developments due to Villamayor and Zelinsky [VZ]. An exposition is given in Chapter II of [M], which we follow here.

Let R be a commutative ring. Idempotents e of R and open–closed subsets U of the spectrum $\text{Spec}(R)$ of R correspond one–one: to e we associate the subset $e^{-1}(1) = \{P \in R | 1 - e \in P\}$; and to U we associate its characteristic function χ_U where $\chi_U = 1 \in R_P$ if and only if $P \in U$. (One has to show that there is a unique idempotent $\chi_U \in R$ with this property.)

Open–closed subsets of $\text{Spec}(R)$ are unions of connected components. Thus is is convenient to pass to the space $\text{Comp}(\text{Spec}(R))$ of connected components of $\text{Spec}(R)$ which we denote $X(R)$ and which is defined by the continuity of the map

$$\text{Spec}(R) \xrightarrow{\quad p \quad} \text{Comp}(\text{Spec}(R)) = X(R).$$

The space $X(R)$ is compact and totally disconnected by construction. It also turns out to be Hausdorff, and hence profinite. The elements of $X(R)$ correspond to the maximal ideals of $B(R)$ and hence $X(R)$ is known as the *Boolean Spectrum* of R.

The points of $X(R)$ are connected components of $\text{Spec}(R)$. If P and Q are prime ideals which lie in the same connected component x then $P \cap B(R) = Q \cap B(R)$; this set of idempotents depends only on the component x. We denote by $I(x)$ the ideal of R which it generates, and we let R_x denote the quotient $R/I(x)$. The ring R_x is *connected*; that is, it has no non-trivial idempotents. For $a \in R$, we let a_x be the coset $a + I(x)$ of a in R_x. Then for an idempotent e we have $e_x = 1$ if and only if $e \equiv 1$ on x.

We generalize this notation to R modules: if M is an R module and $x \in X(R)$ then $M_x = R_x \otimes_R M$.

In fact, the rings R_x are the stalks of a sheaf on $X(R)$. We consider the map of ringed spaces induced by the map p above:

$$(\text{Spec}(R), \mathcal{O}_R) \to (X(R), p_*(\mathcal{O}_R)).$$

Then the stalk $p_*(\mathcal{O}_R)_{X(R),x}$ can be identified with R_x.

The spaces $X(\cdot)$ are functorial: if $f : R \to S$ is a homomorphism of commutative rings then there is an induced continuous function $X(f) : X(S) \to X(R)$ which sends $y \in X(S)$ to the component $X(f)(y)$ of $\text{Spec}(R)$ containing $\{f^{-1}(P) | P \in y\}$. If $X(f)(y) = x$ then f induces a homomorphism $R_x \to S_y$ by $a + I(x) \mapsto f(a) + I(y)$.

As a consequence, the natural maps into a tensor product

$$
\begin{array}{ccc}
S & \longrightarrow & S \otimes_R T \\
\uparrow & & \uparrow \\
R & \longrightarrow & T
\end{array}
$$

induce a commutative diagram of Boolean spectra

$$
\begin{array}{ccc}
X(S) & \longleftarrow & X(S \otimes_R T) \\
\downarrow & & \downarrow \\
X(R) & \longleftarrow & X(T)
\end{array}
$$

and hence a natural continuous map

$$X(S \otimes_R T) \to X(S) \times_{X(R)} X(T).$$

We also need to recall the properties of rings of integer valued functions on profinite spaces. If X is a profinite space, we let $C(X, \mathbb{Z})$ denote the ring of integer valued functions on X. For $x \in X$, the kernel of evaluation at x is a prime ideal $P_x \in \mathrm{Spec}(C(X, \mathbb{Z}))$ which lies in some connected component C_x of this spectrum. The map $x \mapsto P_x \mapsto C_x$ then induces a homeomorphism $X \to X(C(X, \mathbb{Z}))$. If $p : X \to Y$ is a continous map then there is an induced ring homomorphism $C(p) : C(Y, \mathbb{Z}) \to C(X, \mathbb{Z})$ given by $f \mapsto f \circ p$, and we have $X(C(p)) = p$ in the sense that the following diagram commutes:

$$
\begin{array}{ccc}
X & \longrightarrow & X(C(X, \mathbb{Z})) \\
p \downarrow & & \downarrow X(C(p)) \\
Y & \longrightarrow & X(C(Y, \mathbb{Z}))
\end{array}
$$

As an application of these concepts, we prove the following (which will be needed in the sequel, and which illustrates the essential techniques):

Lemma. *Let R be a commutative ring and let W be a profinite space. Then the natural map*

$$X(R \otimes_{\mathbb{Z}} C(W, \mathbb{Z})) \to X(R) \times W$$

is a homeomorphism.

Proof. We first consider the case where W is finite of cardinality w. Then $C(W, \mathbb{Z}) = \mathbb{Z} \times \cdots \times \mathbb{Z} = \mathbb{Z}^{(w)}$ so $R \otimes C(W, \mathbb{Z}) = R^{(w)}$. Thus $X(R \otimes C(W, \mathbb{Z})) = X(R)^{(w)} = W \times X(R)$. Thus we have the result for finite W. In general, we have $W = \mathrm{proj\,lim} W_i$ where W_i is finite. Then $C(W, \mathbb{Z}) = \mathrm{dir\,lim} C(W_i, \mathbb{Z})$ so $R \otimes C(W, \mathbb{Z}) = \mathrm{dir\,lim}(R \otimes C(W_i, \mathbb{Z}))$. Thus $X(R \otimes C(W, \mathbb{Z})) = \mathrm{proj\,lim}(X(R \otimes C(W_i, \mathbb{Z})) = \mathrm{proj\,lim}(X(R) \times W_i) = X(R) \times \mathrm{proj\,lim} W_i = X(R) \times W$, where the third equality follows from the finite case already considered.

Finally, we note that if R and S are commutative rings then $B(R \times S) = B(R) \times B(S)$ from which it follows that $X(R \times S) = X(R) \times X(S)$.

The Adjunction Theorem

Using the notations and results recalled in the preceding section, we can now state the main result.

Theorem 1. *Let R be a commutative ring, Y a profinite topological space, and $\pi : Y \to X(R)$ a continuous surjection. Then there is an R algebra $T_R(Y) = T(Y)$ such that:*

(1) *There is a homeomorphism $Y \to X(T(Y))$ such that*

$$Y \to X(T(Y)) \to X(R) = Y \xrightarrow{\pi} X(R);$$

(2) *If S is any R algebra and $h : X(S) \to Y$ is a continuous map such that*

$$X(S) \xrightarrow{h} Y \xrightarrow{\pi} X(R) = X(S) \to X(R)$$

then there is a unique R algebra map $T(Y) \to S$ such that

$$X(S) \xrightarrow{h} Y \to X(T(Y)) = X(S) \to X(T(Y));$$

(3) *$T(Y) = \cup(Re_1^{(a_1)} \times \ldots Re_n^{(a_n)})$ (directed union) where the $e_1, \ldots, e_n$ are partitions of unity ($e_i e_j = \delta_{ij}$ and $\sum e_i = 1$).*

We begin the proof with the construction of a canonical map from the ring of functions on the Boolean spectrum to the ring:

Proposition 2. *Let R be a commutative ring and define*

$$\Phi_R = \Phi : C(X(R), \mathbb{Z}) \to R$$

by

$$\Phi(f) = \sum_{k \in \mathbb{Z}} k \chi_{f^{-1}(k)}.$$

Then Φ is a ring homomorphism which induces a Boolean algebra isomorphism $B(C(X(R), \mathbb{Z})) \to B(R)$ and $X(\Phi) : X(R) \to X(C(X(R), \mathbb{Z})) = X(R)$ is the identity. The image of Φ is the subring $\mathbb{Z}B(R) = \sum_{e \in B(R)} \mathbb{Z}e$.

Also, Φ is functorial in R: if $\alpha : R \to S$ is a commutative ring homomorphism then the following diagram commutes:

$$
\begin{array}{ccc}
C(X(R), \mathbb{Z}) & \xrightarrow{\Phi_R} & R \\
{\scriptstyle C(X(\alpha))}\big\downarrow & & \big\downarrow{\scriptstyle \alpha} \\
C(X(S), \mathbb{Z}) & \xrightarrow[\Phi_S]{} & S.
\end{array}
$$

Proof. A continuous function $f : X(R) \to \mathbb{Z}$ has finite image, so the sum in the defintion of Φ is actually finite, as are the sums which will appear in this proof. Moreover, the various $f^{-1}(k)$'s are disjoint. The characteristic function of an intersection is the product of characteristic functions: $\chi_{U \cap V} = \chi_U \cdot \chi_V$. And if $U \cap V = \emptyset$, then the characteristic function $\chi_{U \cup V}$ of the union is the sum $\chi_U + \chi_V$, with a similar result for sets of mutally disjoint sets. Let f, g belong to $C(X(R), \mathbb{Z})$. For simplicity in the following calculations, we will denote the characteristic function of $f^{-1}(i)$ by χ_i and the characteristic function of $g^{-1}(i)$ by ψ_i. Then $\sum_i \chi_i = \sum_i \psi_i = 1$ so

$$\Phi(f) = \sum_k k\chi_k = \sum_k \sum_i k\chi_k\psi_i$$

and

$$\Phi(g) = \sum_m m\psi_m = \sum_m \sum_j m\psi_m\chi_j$$

so

$$\Phi(f) + \Phi(g) = \sum_{p,q}(p+q)\chi_p\psi_q.$$

Since $(f+g)^{-1}(k) = \cup_{p+q=k}(f^{-1}(p) \cap g^{-1}(q))$, its characteristic function is

$$\sum_{p+q=k} \chi_p\psi_q$$

so that

$$\Phi(f+g) = \sum k \sum_{p+q=k} \chi_p\psi_q = \Phi(f) + \Phi(g).$$

Similar reasoning applies to the product fg: here we have to use the fact that $(\chi_k\psi_i)\cdot(\chi_j\psi_m)$ is the characteristic function of $(f^{-1}(k)\cap g^{-1}(i))\cap(f^{-1}(m)\cap g^{-1}(j))$ and hence is zero unless $k = j$ and $i = m$. Thus we find that

$$\Phi(f) \cdot \Phi(g) = \sum_{p,q}(pq)\chi_p psi_q.$$

Now $(fg)^{-1}(k) = \cup_{pq=k}(f^{-1}(p) \cap g^{-1}(q))$ so that its characteristic function is $\sum_{pq=k} \chi_p\psi_q$ and so that

$$\Phi(fg) = \sum k \sum_{pq=k} \chi_p\psi_q = \Phi(f) \cdot \Phi(g).$$

This establishes that Φ is a ring homomorphism. If e is an idempotent of R, U is the corresponding open–closed subset of $X(R)$ ($U = \{x \in X(R)|e_x = 1\}$) and f is the characteristic function of U then $\Phi(f) = e$, so Φ is surjective on idempotents. Idempotents of $C(X(R),\mathbb{Z})$ are characteristic functions of open–closed subsets of $X(R)$. If V is such a subset and f its characteristic function then $\Phi(f) = \chi_V$ is the idempotent of R with $V = \{x \in X(R)|\Phi(f)_x = 1\}$. So if $\Phi(f) = 0$ then $V = \emptyset$ so $f = 0$. Thus Φ is injective on idempotents and induces a Boolean alegbra isomorphism $B(C(X(R),\mathbb{Z})) \to B(R)$ and hence a homeomorphism on Boolean spectra, which then induces the claimed identity map on $X(R)$. The image of Φ is clearly $\mathbb{Z}B(R)$.

Finally, suppose we have a commutative ring homomorphism $\alpha : R \to S$, and let $e \in B(R)$ with $\Phi_R(f) = e$. Then for a component $x \in X(R)$, and a prime ideal $P \in x$, $f(x) = 1$ if and only if $1 - e \in P$. Suppose $y \in X(S)$ is such that

$X(\alpha)(y) = x$ (there may not be any such y for a given x) and suppose $Q \in y$ is a prime ideal; then $\alpha^{-1}(Q)$ is a prime ideal in x. So $C(X(\alpha))(f)(y) = f(X(\alpha)(y)) = 1$ if and only if $1 - e \in \alpha^{-1}(Q)$ if and only if $1 - \alpha(e) \in Q$; which then implies that $\Phi_S(C(X(\alpha))(f)) = \alpha(\Phi_R(f))$, and hence the necessary diagram commutes.

Proposition 2 shows that every commutative ring R is an algebra over the ring $C(X(R), \mathbb{Z})$ of continuous integer valued functions on its Boolean spectrum. If $Y \to X(R)$ is a continuous surjection of profinite spaces, then $C(Y, \mathbb{Z})$ is also an algebra over $C(X(R), \mathbb{Z})$. We tensor the two to produce the algebra $T_R(Y)$ of Theorem 1:

Definition 3. Let R be a commutative ring, Y a profinite space, and $\pi : Y \to X(R)$ a continuous surjection. Then

$$T_R(Y) = R \otimes_{C(X(R), \mathbb{Z})} C(Y, \mathbb{Z})$$

where R is a $C(X(R), \mathbb{Z})$ algebra by Φ_R and $C(Y, \mathbb{Z})$ is a $C(X(R), \mathbb{Z})$ algebra by $C(\pi)$.

We need to analyze futher the maps $\Phi = \Phi_R$ and $C(\pi)$ of Definition 3. First, suppose $x \in X(R)$. We identify $X(R)$ and $X(C(X(R), \mathbb{Z}))$ and consider x in the latter also. Then Φ induces a map

$$\mathbb{Z} = C(X(R), \mathbb{Z})_x \to R_x$$

(in the first equality $\mathbb{Z}$ maps in as the constant functions).

The notation R_x is potentially ambiguous: it could mean R modulo the ideal generated by the idempotents in R which are 0 on $x \in X(R)$, or it could stand for the tensor product

$$C(X(R), \mathbb{Z})_x \otimes_{C(X(R), \mathbb{Z})} R.$$

In fact, since every idempotent of R lies in the image of Φ these are the same ring.

We also consider the ring $C(Y, \mathbb{Z})_x$, which is $C(Y, \mathbb{Z})$ modulo the ideal generated by the $C(\pi)$ image of the idempotents of $C(X(R), \mathbb{Z})$ which are 0 at x. Of course these images are the idempotents of $C(Y, \mathbb{Z})$ which are constant on the fibres of π and which vanish on $\pi^{-1}(x)$, and these in turn generate the kernel of the (surjective) restriction map $C(Y, \mathbb{Z}) \to C(\pi^{-1}(x), \mathbb{Z})$.

As an application of these remarks we note:

Lemma 4. Let R be a commutative ring, Y a profinite space, $\pi : Y \to X(R)$ a continuous surjection, and $x \in X(R)$. Then

$$X(T_R(Y)_x) = \pi^{-1}(x) \subseteq Y.$$

Proof. Let $C = C(X(R), \mathbb{Z})$. We have

$$T(Y)_x = R_x \otimes (R \otimes_C C(Y, \mathbb{Z}) = R_x \otimes_{C_x} C(Y, \mathbb{Z})_x = R_x \otimes_{\mathbb{Z}} C(\pi^{-1}(x), \mathbb{Z})$$

and then the Lemma above implies that $X(T(Y)_x) = X(R_x) \times \pi^{-1}(x)$, but R_x has no non–trivial idempotents so $X(R_x)$ is a point.

We can now prove point (1) of Theorem 1:

Proof of Thm 1.1. Let let $X = X(R)$ and we let $x \in X$. Then we have by Lemma 4 a homeomorphism $X(T(Y)_x) \to \pi^{-1}(x)$. This homeomorphism is compatible with the canonical projection $h : X(T(Y)) \to Y$ which comes from the homeomorphism $X(C(Y,\mathbb{Z}) \to Y$ following the projection $X(T(Y)) \to X(C(Y,\mathbb{Z}))$, the latter coming from the definition of $T(Y)$ as a tensor product, the second factor of which is $C(Y,\mathbb{Z})$. These two maps fit into a commutative diagram:

$$
\begin{array}{ccc}
X(T(Y)) & \xrightarrow{\ h\ } & Y \\
\uparrow & & \uparrow \\
X(T(Y)_x) & \xrightarrow{\ \simeq\ } & \pi^{-1}(y).
\end{array}
$$

This shows that h is surjective. Suppose that $z, w \in X(T(Y))$ and that $h(z) = h(w)$. Let $p_1 : X(T(Y)) \to X$ be defined by inclusion of the first factor in the tensor product. Then we have the commutative diagram

$$
\begin{array}{ccc}
X(T(Y)) & \xrightarrow{\ h\ } & Y \\
\downarrow & & \downarrow{\scriptstyle \pi} \\
X & \xrightarrow{\ \text{id}\ } & X
\end{array}
$$

which shows that $p_1(z) = \pi(h(z)) = \pi(h(w)) = p_1(w)$; if x is this element of X then both z and w lie in $\pi^{-1}(x)$ and hence the first diagram shows $z = w$ so h is injective. Thus h is a homeomorphism.

A similar argument proves point (2) of Theorem 1:

Proof of Thm 1.2. The map $h : X(S) \to Y$ defines a homomorphism $C(h) : C(Y,\mathbb{Z}) \to C(X(S),\mathbb{Z})$, and the assumption on h implies that

$$
C(X(R),\mathbb{Z}) \to C(Y,\mathbb{Z}) \xrightarrow{\ C(h)\ } C(X(S),\mathbb{Z}) = C(X(R),\mathbb{Z}) \to C(X(S),\mathbb{Z}).
$$

Then the functorality of the map Φ implies that

$$
\begin{array}{ccc}
C(Y,\mathbb{Z}) & \xrightarrow{\ \Phi_S C(h)\ } & S \\
\uparrow & & \uparrow \\
C(X(R),\mathbb{Z}) & \xrightarrow[\ \Phi_R\ ]{} & R
\end{array}
$$

commutes, and this diagram yields the map $T(Y) \to S$ realizing h. It is clear that this map is unique.

To establish point (3) of Theorem 1, we need to analyze the construction of $T(Y)$.

Proof of Thm 1.3. Let $X = X(R)$, and let $\{X_i, f_{im} : X_m \to X_i | i, m \in I\}$ be an inverse system of finite sets with surjective transition maps whose inverse limit is

X. We can choose a compatible inverse system $\{Y_j, g_{jn} : Y_n \to Y_j | j, n \in J\}$, also with surjective transition maps, whose inverse limit is Y: that is, there is a map $\phi : J \to I$ such that Y_j maps to $X_{\phi(j)}$ and that the inverse limit of these maps is π. By setting $f_{pq} =$ id when $\phi(p) = \phi(q)$ for $q > p$ elements of J we can also regard X as an inverse limit over J. Then we have

$$C(Y, \mathbb{Z}) = \text{proj } \lim_J C(Y_j, \mathbb{Z}) \supseteq \text{proj } \lim_J C(X_j, \mathbb{Z})$$

so that

$$T(Y) = \text{dir } \lim_J (R \otimes_{C(X_j, \mathbb{Z})} C(Y_j, \mathbb{Z})).$$

We note that the direct limit in the above equation is actually a directed union; this follows from the fact that we used inverse sustems with surjective transition maps.

Now if $X_j = \{x_1, \ldots, x_n\}$ has cardinality n, the fibres of $X \to X_j$ have the idempotents $e_1, \ldots, e_n$ as their characteristic functions, and the fibre of $Y_j \to X_j$ over e_i has cardinality a_i, then

$$R \otimes_{C(X_j, \mathbb{Z})} C(Y_j, \mathbb{Z}) = Re_1^{(a_1)} \times \cdots \times Re_n^{(a_n)}.$$

This establishes the direct limit formula of Theorem 1.3. The final assertion, namely that $T(Y)$ is faihfully flat over R, will follow from noting that each of the terms in the directed union is R projective and faithful.

APPLICATIONS

The fact that every Boolean algebra can be represented as the Boolean algebra of open–closed subsets of some profinite space (this is the Stone Representation Theorem) means that an immediate corollary of Theorem 1 is as the following Boolean algebra extension results:

Theorem 5. *Let R be a commutative ring B a Boolean ring, and let $B \subseteq B(R)$ be a Boolean algebra extension. Then there is an R algebra R_B such that:*

(1) *There is a Boolean algebra isomorphism $B(R_B) \to B$ such that*

$$B(R) \subseteq B(R_B) \to B = B(R) \subseteq B;$$

(2) *If S is any R algebra and $h : B \to B(S)$ is a Boolean algebra homomorphism such that*

$$B(R) \subseteq B \to B(S) = B(R) \subseteq B(S)$$

then there is a unique R algebra map $R_B \to S$ such that

$$B(R_B) \to B \to B(S) = B(R_B) \to B(S) :$$

(3) *R_B is generated over R by $B(R_B)$ and R_B is faithfully flat over R.*

Proof. We take a profinite space Y with $B(C(Y, \mathbb{Z})$ isomorphic to B and then let $R_B = T_R(Y)$; the properties (1), (2), and (3) are then resatements of (or in the case of (3), consequences of) the corresponding properties of Theorem 1.

Thus Theorem 1 answers the question posed in the introduction, as well as explaining how, for example, to adjoin idempotents to a commutative ring to make its Boolean algebra of idempotents complete.

The author's original interest in this question was in regards to adjoing idempotents to a ring R in order to make its Boolean algebra of idempotents injective in the category of Boolean algebras; this is a necessary step in the separable Galois theory of commuative rings with idempotents, [M, Thm. IV.20, p.93]. One wants to do this in a minimal fashion, which is easiest to explain in terms of the space $X(R)$: the idea is to construct a minimal projective cover $Y \rightarrow X(R)$, where minimal means in the sense of Gleason cover, [M, Defn. I.22, p.13]. This lead to a need for a result like Theorem 1 parts 1 and 3. In fact this was solved [M Prop IV.12, p.86] by means of a more complicated construction than that described here, and without the universal property of Theorem 1 part 2. Of course, after the universal property has been established, we see that the construction must coincide with the ring $T(Y)$ produced here, and if fact the notation was chosen with that in mind.

REFERENCES

[M] Magid, A., *The Separable Galois Theory of Commutative Rings*, Marcel Dekker, Inc., New York, 1974.

[P] Pierce, R., *Modules over commutative regular rings*, Mem. Amer. Math. Soc. No. 70, Amer. Math. Soc., Providence, 1967.

[VZ] Villamayor, O. and Zelinsky, D., *Galois theory with infinitely many idempotents*, Nagoya Math. J. **35** (1969), 83–98.

Separable Polynomials and Weak Henselizations

THOMAS MCKENZIE Department of Mathematics, Bradley University, Peoria, IL 61625

1: Introduction

Throughout this paper we will assume that all rings are commutative rings with identity, that ring homomorphisms preserve identities, and that a ring and its subrings have the same identity. In the second section of this paper we list some results which will be used in the sequel. We then consider some of the constructions and results of classical algebraic number theory in the context of the infinite Galois theory of rings. Most of the constructions and results in this second section either have appeared in the literature or are modifications of classical results.

In the third section of this paper we study the separable polynomials over a ring R. A monic polynomial $f \in R[x]$ is defined to be indecomposable if whenever there exist monic polynomials $g, h \in R[x]$ such that $f = gh$ it follows that $g = 1$ or $h = 1$. A theorem in [HM] states that every separable polynomial over a ring can be factored uniquely into a product of indecomposable separable polynomials if and only if the ring has exactly two idempotents. In section three we expand upon this result.

By a local ring we mean a (not necessarily Noetherian) ring with a unique maximal ideal. In section four we assume that R is a local ring. We then construct a local ring $R^{\#}$ which, among other things, both contains R as a subring and has a local separable closure. The ring $R^{\#}$ is called a weak Henselization of R. We then consider the separable polynomials over $R^{\#}$. In particular we show that if f is a monic polynomial over $R^{\#}$ then f is an indecomposable separable polynomial over $R^{\#}$ if and only if f is irreducible and separable when viewed as a polynomial over the residue class field of $R^{\#}$. We also prove that if the residue class field of R is infinite then $R = R^{\#}$ if and only if every indecomposable separable polynomial over R is irreducible when viewed as a polynomial over the residue class field of R. Later we consider some of the properties of the R-algebra $R^{\#}$. In particular we prove $R^{\#}$ is a canonical object.

Local rings which are integrally closed in their quotient fields are studied in the last section. The set of monic polynomials over a Henselization of such a ring has been the object of much study. One of the ideas in this paper is that the set of separable polynomials over $R^{\#}$ has properties which are analogous to the properties of the set of monic polynomials over a Henselization of R. In section five we study the relationship between $R^{\#}$ and a Henselization of R. We show that $R^{\#}$ can be embedded in any Henselization of R. Next we find a necessary and sufficient condition for this embedding to be an isomorphism. We conclude section five by proving that if R is the ring of integers localized at the prime ideal which is generated by 3 then $R^{\#}$ is not isomorphic to any Henselization of R.

A few words about definitions which we will use throughout this paper. If S is a ring and T is a ring extension of S then T will be called a finite extension of S if it is finitely generated as an S-module. We define $Aut_S T$ to be the ring automorphisms of T which leave fixed the elements in S. If G is a group of ring automorphisms of T then we let T^G denote the set of elements in T which are left fixed by every element in G.

2: Maximal ideals

A ring is said to be connected if 0 and 1 are the only idempotents in the ring. Throughout this section we will assume that R and S are connected rings and that S is a locally separable extension of R. In this section we consider the action of the $Aut_R S$ on the maximal ideals of S. Our main goal is to list a number of results which will be used later in this paper.

If M' is a maximal ideal of S then $M' \cap R$ is a maximal ideal of R since S is an integral extension of R. If M is a maximal ideal of R and M' is a maximal ideal of S such that $M' \cap R = M$ we say that M' lies over M and we write $f(M'|M)$ for the degree of the field S/M' over the field R/M (this could be infinite).

THEOREM 2.1   If M is a maximal ideal of R and X is the set of maximal ideals of S which lie over M then $\cap_{Q \in X} Q = MS$.

Proof: First assume that S is a finite R-algebra. Let

$$M_1, \ldots, M_t$$

be the maximal ideals of S which lie over M. It is clear that $MS \subseteq M_1 \cap \ldots \cap M_t$. To establish the reverse inclusion note that by Theorem 7.1 on page 72 in [DI] $S/(MS)$ is a separable R/M-algebra. Thus $S/(MS)$ is a direct sum of a finite number of finite dimensional, separable field extensions of R/M. So the Jacobson radical of $S/(MS)$ is zero. Hence $M_1 \cap \ldots \cap M_t \subseteq MS$.

Now assume that S is not necessarily a finite R-algebra. Again it is clear that $MS \subseteq \cap_{Q \in X} Q$. To establish the reverse inclusion let $\alpha \in \cap_{Q \in X} Q$. Let T be a finite projective separable extension of R such that T is contained in S and T contains α. By the argument above $\alpha \in MT$. Thus $\alpha \in MS$. Hence $\cap_{Q \in X} Q \subseteq MS$. This completes the proof.

Let M be a maximal ideal of R and assume for a moment that S is a finite normal extension of R. Note that there are only finitely many maximal ideals of S which lie over M. Let M_1 and M_2 be two of the maximal ideals of S which lie over M. If $m_1 \in M_1$ then

$$\prod_{\sigma \in Aut_R S} \sigma(m_1) \in M_1 \cap R = M \subseteq M_2.$$

So there must exist $\sigma \in Aut_R S$ such that $\sigma(m_1) \in M_2$. It follows that

$$M_1 \subseteq \bigcup_{\sigma \in Aut_R S} \sigma(M_2).$$

Thus there exists $\sigma \in Aut_R S$ such that $M_1 = \sigma(M_2)$. From this we see that if S is a finite, normal extension of R then the maximal ideals of S which lie over a fixed maximal ideal of R are permuted transitively by $Aut_R S$. In fact, this is true even if the rank of S over R is not necessarily finite.

LEMMA 2.2   If S is a normal extension of R, M is a maximal ideal of R, and M_1 and M_2 are maximal ideals of S which lie over M then there exists $\sigma \in Aut_R S$ such that $\sigma(M_1) = M_2$.

Proof: If X is a finite subset of S then there exists a finite Galois extension T of R such that $X \subseteq T$, T is a subring of S and R is a subring of T. So S is the union of the finite Galois extensions of R which are contained in S. Let $(T_\alpha)_{\alpha \in I}$ be the set of all finite Galois extensions of R which are contained in S. For $\alpha \in I$ let

$$H_\alpha = \{\sigma \in Aut_R(S) | \sigma(M_1 \cap T_\alpha) = M_2 \cap T_\alpha\}.$$

We wish to show that $\bigcap_{\alpha \in I} H_\alpha \neq \emptyset$. We begin by showing that each H_α is closed in $Aut_R S$.

    Let $\alpha \in I$. Since T_α is finite over R, $Aut_{T_\alpha} S$ is open in $Aut_R S$. Let $\gamma \in Aut_R S - H_\alpha$. Then for all $\tau \in Aut_{T_\alpha} S$,

$$\gamma[\tau(M_1 \cap T_\alpha)] = \gamma(M_1 \cap T_\alpha) \neq M_2 \cap T_\alpha.$$

So for every $\gamma \in Aut_R S - H_\alpha$ there exists an open set in $Aut_R S$ which contains γ and is contained in $Aut_R S - H_\alpha$. Hence $Aut_R S - H_\alpha$ is open and H_α is closed.

    In order to show that $\bigcap_{\alpha \in I} H_\alpha \neq \emptyset$ we now choose a finite subset J of I. We will show that $\bigcap_{\alpha \in J} H_\alpha \neq \emptyset$. Note that there exists $\beta \in I$ such that $T_\alpha \subseteq T_\beta, \forall \alpha \in J$. If $\gamma \in Aut_R S$ such that $\gamma(M_1 \cap T_\beta) = M_2 \cap T_\beta$ then $\gamma(M_1 \cap T_\alpha) = M_2 \cap T_\alpha$ for all $\alpha \in J$. Thus $H_\beta \subseteq H_\alpha$ for all $\alpha \in J$. Since T_β is finite we know from the discussion preceeding the statement of this lemma that $H_\beta \neq \emptyset$. Thus $\bigcap_{\alpha \in J} H_\alpha \neq \emptyset$.

    We have shown that each H_α is closed and that $\bigcap_{\alpha \in J} H_\alpha \neq \emptyset$ for every finite subset J of I. Since $Aut_R S$ is compact the finite intersection property yields $\bigcap_{\alpha \in I} H_\alpha \neq \emptyset$. Now let $\gamma \in \bigcap_{\alpha \in I} H_\alpha$, let $m_1 \in M_1$, and let $\alpha \in I$ such that $m_1 \in T_\alpha$. Note that $\sigma(m_1) \in T_\alpha \cap M_2$. Hence $\sigma(M_1) = M_2$. This completes the proof.

    As we noted above if M is a maximal ideal of R and S is a finite extension of R then there are only finitely many, say g, maximal ideals of S which lie over M and by the definition of rank we have

$$rank_R(S) = \sum_{i=1}^{g} f(M_i | M).$$

If S is a normal extension of R then Lemma 2.2 implies that $S/M_i \simeq S/M_j$, for all $i, j \in \{1, \ldots, g\}$. Thus

$$f(M_i | M) = f(M_j | M), \forall i, j \in \{1, \ldots, g\}$$

and hence

$$rank_R S = g \cdot f(M_i|M), \forall i \in \{1, \ldots, g\}.$$

If T is an extension of R and M' is a maximal ideal of T then we define the decomposition group of M' (with respect to $Aut_R T$) to be

$$\{\sigma \in Aut_R T | \sigma(M') = M'\}$$

and the inertial subgroup of M' (with respect to $Aut_R T$) to be

$$\{\sigma \in Aut_R T | \sigma(\alpha) - \alpha \in M', \forall \alpha \in T\}.$$

We write $D(M')$ for the decomposition group of M' and $E(M')$ for the inertial subgroup of M'.

THEOREM 2.3   If T is a connected, normal extension of R and $Aut_R T$ is finite then the following are equivalent:

(i) T is a separable R-algebra;

(ii) if M' is a maximal ideal of T then $E(M')$ is trivial.

Proof: This follows from Proposition 1.2 on page 80 in [DI].

CORROLLARY 2.4   If T is a connected, normal, locally separable extension of R then $E(M')$ is trivial for every maximal ideal M' of T.

Proof: If T is finite over R then by Theorem 2.3 $E(M') = \{1\}$ for every maximal ideal M' of T. Now let us consider the case where T is not necessarily finite over R. Assume by way of contradiction that there exists a maximal ideal M' of T such that $E(M')$ is not trivial. Then there exists $1 \neq \sigma \in Aut_R T$ such that $\sigma(x) - x \in M', \forall x \in T$. Let $t \in T$ such that $\sigma(t) \neq t$ and let T_0 be a finite Galois extension of R such that $t \in T_0$ and $T_0 \subseteq T$. By Theorem 2.3, there exists $t_0 \in T_0$ such that $\sigma(t_0) - t_0 \notin T_0 \cap M'$. This is a contradiction. Hence $E(M')$ is trivial.

We now use the argument that (i) implies (ii) in the proof of Theorem 2.3 to prove the following theorem.

THEOREM 2.5   Assume that T is a connected separable extension of R. If M' is a maximal ideal of T then $E(M')$ is trivial.

Proof: Let $\sigma \in G$ and note that $\overline{\sigma} : T \otimes_R T \to T \otimes_R T$ given by $\overline{\sigma}(x \otimes y) = x \otimes \sigma(y)$ is a T-algebra automorphism of $T \otimes_R T$. First we show that there exist $x_1, \ldots, x_n, y_1, \ldots, y_n \in T$ such that $\delta_{\sigma,1} = \sum_{i=1}^n x_i \sigma(y_i)$ where 1 is the identity in $Aut_R T$ and

$$\delta_{\sigma,1} = \begin{cases} 1, & \text{if } \sigma = 1; \\ 0 & \text{if } \sigma \neq 1. \end{cases}$$

Let e be a separability idempotent for T and let $\mu : T \otimes_R T \to T$ be the R-algebra homomorphism given by $\mu(x \otimes y) = x \cdot y$. For $x \in T$,

$$\begin{aligned} x \cdot \mu((\overline{\sigma}(e)) &= \mu(x \otimes 1 \cdot \overline{\sigma}(e)) \\ &= \mu(\overline{\sigma}(x \otimes 1 \cdot e)) \\ &= \mu(\overline{\sigma}(1 \otimes x \cdot e)) \quad \text{(since } e \text{ is a separability idempotent)} \\ &= \sigma(x) \cdot \mu(\overline{\sigma}(e)). \end{aligned}$$

Note that $\mu(\overline{\sigma}(e))$ is an idempotent in T. Thus $\mu(\overline{\sigma}(e))$ is equal to 0 or 1. If $\sigma = 1$ then by the definition of e, $\mu(\overline{\sigma}(e)) = 1$. On the other hand, if $\sigma \neq 1$ then there exists $x \in T$ such that $\sigma(x) \neq x$ and the equality $x \cdot \mu((\overline{\sigma}(e)) = \sigma(x) \cdot \mu(\overline{\sigma}(e))$ yields $\mu(\overline{\sigma}(e)) = 0$. Thus if $e = \sum_{i=1}^{n} x_i \otimes y_i$ then $\delta_{\sigma,1} = \sum_{i=1}^{n} x_i \sigma(y_i)$.

We now complete the proof of the theorem. Assume by way of contradiction that $E(M')$ is not trivial. Then there exists $\sigma \neq 1 \in Aut_R T$ such that $(1 - \sigma)T \subseteq M'$. It follows that $1 = \sum_{i=1}^{n} x_i(y_i - \sigma(y_i))$ is in M'. This is a contradiction. Hence $E(M')$ is trivial. This completes the proof.

We now turn our attention to the fixed ring of the decomposition group.

THEOREM 2.6 If S is a finite Galois extension of R, M is a maximal ideal of R, and M' is a maximal ideal of S such that $M' \cap R = M$ then:

(i) $rank_R S^{D(M')}$ is equal to the number of distinct maximal ideals of S which lie over M;

(ii) $rank_{S^{D(M')}} S = f(M'|M)$.

Proof: (i) Let $G = Aut_R S$ and let M'' be a maximal ideal of S which lies over M. By Lemma 2.2 there exists $\gamma \in Aut_R S$ such that $\gamma(M') = M''$. If $\delta \in D(M')$ then $(\gamma \cdot \delta)(M') = \gamma(M') = M''$. So there is a bijective correspondence between the distinct cosets of $G/D(M')$ and the maximal ideals of S which lie over M. By Theorem 4.6 on page 107 of [DI], $rank_R S^{D(M')} = |G/D(M')|$. Hence $rank_R S^{D(M')}$ is equal to the number of distinct maximal ideals of S which lie over M.

(ii) Note that $rank_R S = rank_{S^{D(M')}} S \cdot rank_R S^{D(M')}$. Let g be the number of distinct maximal ideals in S lying over M. Then $rank_R S = g \cdot f(M'|M)$ and by part (i) of this theorem, $rank_R S^{D(M')} = g$. Thus $g \cdot f(M'|M) = rank_{S^{D(M')}} S \cdot g$. Hence $rank_{S^{D(M')}} S = f(M'|M)$.

Next we use the preceding theorem to show that the decomposition group is isomorphic to the automorphism group of an extension of fields.

THEOREM 2.7 If S is a normal extension of R, M is a maximal ideal of R, and M' is a maximal ideal of S such that $M' \cap R = M$ then $D(M') \simeq Aut_{R/M} S/M'$.

Proof: Let Δ be the group homomorphism from $D(M')$ to $Aut_{R/M} S/M'$ given by $\overline{\sigma}(s + M') = \sigma(s) + M'$. By Corollary 2.4,

$$\{\sigma \in D(M') | \sigma(x) - x \in M', \forall x \in S\} = \{1\}.$$

Thus Δ is an injective group homomorphism. If S is finite over R then by Theorem 2.6 part (i)

$$|D(M')| = f(M'|M)$$

and hence Δ is a group isomorphism.

To prove that Δ is surjective when S is not necessarily finite over R let

$$\tau \in Aut_{R/M} S/M'.$$

We know that $S = \cup_{\alpha \in I} T_\alpha$ where each T_α is a finite Galois extension of R which is contained in S. For $\alpha \in I$ let $M'_\alpha = M' \cap T_\alpha$ and let

$$E_\alpha = \{\sigma \in Aut_R S | \sigma|_{T_\alpha} \in D(M'_\alpha) \text{ and } \Delta(\sigma)|_{T_\alpha/M'_\alpha} = \tau|_{T_\alpha/M'_\alpha}\}.$$

We wish to show that $\bigcap_{\alpha \in I} E_\alpha \neq \emptyset$. We begin by showing that each E_α is closed in $Aut_R S$.

Let $\alpha \in I$. Since T_α is finite, $Aut_{T_\alpha} S$ is open in $Aut_R S$. If $\gamma \in Aut_R S - E_\alpha$ then $\gamma \circ \eta \in Aut_R S - E_\alpha$ for all $\eta \in Aut_{T_\alpha} S$. Thus there exists an open subset of $Aut_R S$ which contains γ and is contained in $Aut_R S - E_\alpha$. Hence $Aut_R S - E_\alpha$ is open and E_α is closed.

In order to show that $\bigcap_{\alpha \in I} E_\alpha \neq \emptyset$ we now choose a finite subset J of I. We will show that $\bigcap_{\alpha \in J} E_\alpha \neq \emptyset$. Note that there exists $\beta \in I$ such that $T_\alpha \subseteq T_\beta$ for all $\alpha \in J$. We know from the discussion above that there exists $\eta \in Aut_R T_\beta$ such that $\Delta(\eta) = \tau|_{T_\beta / M'_\beta}$. Let $\tilde{\eta} \in Aut_R S$ such that $\tilde{\eta}|_{T_\alpha} = \eta$. Note that $\tilde{\eta} \in E_\alpha$ for all $\alpha \in J$. Thus $\bigcap_{\alpha \in J} E_\alpha \neq \emptyset$.

We have shown that each E_α is closed and that $\bigcap_{\alpha \in J} E_\alpha \neq \emptyset$ for every finite subset J of I. Since $Aut_R S$ is compact the finite intersection property yields $\bigcap_{\alpha \in I} E_\alpha \neq \emptyset$. Let $\eta \in \bigcap_{\alpha \in I} E_\alpha$. Note that $\Delta(\eta) = \tau$. Hence Δ is surjective. This completes the proof.

Finally, we compute the inertial degree of the fixed ring of the decomposition group.

THEOREM 2.8   If S is a normal extension of R, M is a maximal ideal of R, M' is a maximal ideal of S such that $M' \cap R = M$, and $M'' = M' \cap S^{D(M')}$ then $f(M''|M) = 1$.

Proof: First assume that S is a finite R-algebra. Lemma 2.2 implies that M' is the only maximal ideal of S which lies over M''. Thus $rank_{S^{D(M')}} S = f(M'|M'')$. It follows from part (ii) of Theorem 2.6 that $f(M'|M'') = f(M'|M)$. Hence $f(M''|M) = 1$.

To prove the general case let $\alpha \in S^{D(M')}$ and let $T \subseteq S$ be a finite Galois extension of R such that $\alpha \in T$. By the preceding paragraph there exists $\beta \in R$ such that $\alpha - \beta \in T^{D(M')} \cap M'$. Hence $R/M \simeq S^{D(M')}/M''$.

3: Separable Polynomials

Throughout this section we will assume that S is a ring and that R is a connected ring. A monic polynomial f in $S[x]$ will be called indecomposable if whenever there exist monic polynomials $g, h \in S[x]$ with $f = g \cdot h$ it follows that $g = 1$ or $h = 1$. The main objects of study in this section are the separable polynomials over a ring. We begin with a result from [HM].

THEOREM 3.1   The following are equivalent:

(i) S is a connected ring;

(ii) If f is a separable in S and $f_1, \ldots, f_n, g_1, \ldots, g_m$ are monic indecomposable polynomials in S such that

$$f = f_1 \cdot \ldots \cdot f_n = g_1 \cdot \ldots \cdot g_m$$

it follows that $n = m$ and for each $i \in \{1, \ldots, n\}$ there exists $j \in \{1, \ldots, m\}$ such that $f_i = g_j$.

(iii) If f is a separable polynomial in S and $f_1, \ldots, f_n, g_1, \ldots, g_m$ are indecomposable separable polynomials in S such that

$$f = f_1 \cdot \ldots \cdot f_n = g_1 \cdot \ldots \cdot g_m$$

it follows that $n = m$ and for each $i \in \{1, \ldots, n\}$ there exists $j \in \{1, \ldots, m\}$ such that $f_i = g_j$.

Proof: This is Theorem 1.6 on page 24 of [HM].

DEFINITION 3.2   If $f, f_1, \ldots, f_n$ are monic polynomials in $R[x]$ such that f is separable; $f_1, \ldots, f_n$ are indecomposable; and $f = f_1 \cdot \ldots \cdot f_n$ then the factorization $f_1 \cdot \ldots \cdot f_n$ is called the unique decomposition of f in $R[x]$.

We now consider an example of the unique decomposition of a separable polynomial over a connected ring. Let $\mathbf{Z}$ denote the ring of integers , $T = \mathbf{Z}/(9\mathbf{Z})$, M' denote the unique maximal ideal of T, and $f \in T[x]$ be given by $f = x^3 + 2x^2 + 3x + 2$. Let $\overline{f}$ denote the polynomial in $(T/M')[x]$ which is obtained by reducing the coefficients of f modulo M'. Note that $2 \cdot \overline{f} + (x^2 + 2x) \cdot \overline{f'} = 1$ and thus by Theorem 1.1 on page 22 of [HM] $\overline{f}$ is separable in $(T/M')[x]$. Since M' is the unique maximal ideal of T, Theorem 2.2 on page 467 of [J] implies that f is separable in $T[x]$. Since T is a connected ring, f has a unique decomposition. Let $f_1, f_2 \in T[x]$ be given by $f_1 = x + 1$, and $f_2 = x^2 + x + 2$. For $i \in \{1, \ldots, 2\}$ let $\overline{f}_i \in (T/M')[x]$ be the polynomial which is obtained by reducing the coefficients of f_i modulo M'. Each $\overline{f}_i$ is irreducible in $(T/M')[x]$ and hence each f_i is indecomposable in $T[x]$. Since $f = f_1 \cdot f_2$ we see that this is the unique decomposition on f in $T[x]$.

If T is an R-algebra and α is an element of T we say that α is a primitive element of the R-algebra T if $T = R[\alpha]$.

THEOREM 3.3   Assume that S is a finite projective connected extension of R and let $\alpha \in S$. The following are equivalent:

(i) There exists a separable, indecomposable polynomial $f \in R[x]$ such that $f(\alpha) = 0$ and $R[x]/(f) \simeq S$;

(ii) α is a primitive element of the R-algebra S

Proof: $(i) \Rightarrow (ii)$. Let $\varphi : R[x] \to R[\alpha]$ be given by $\varphi(g) = g(\alpha)$. It is clear that $f \in ker\varphi$. Let $g \in ker\varphi$. By Theorem 4.4 on page 111 of [DI] there exist $\alpha_1, \ldots, \alpha_m \in \Omega_R$ such that $f = (x - \alpha_1) \cdot \ldots \cdot (x - \alpha_m)$ and $\alpha_i - \alpha_j$ is a unit in Ω_R for all $i, j \in \{1, \ldots, m\}$ with $i \neq j$. Note that $g(\alpha_i) = 0, \forall i \in \{1, \ldots, n\}$. So by Corollary 2.8 on page 470 of [J] $g = f \cdot h$ for some $h \in R[x]$. Thus $(f) = ker\varphi$. Hence $R[x]/(f)$ is isomorphic to $R[\alpha]$ and $R[\alpha]$ is a separable extension of R with $rank_R R[\alpha] = rank_R S$. Hence by Theorem 3.6 on page 107 of [DI] $R[\alpha] = S$.

$(ii) \Rightarrow (i)$ Let $\alpha_1, \ldots, \alpha_n$ be the distinct elements in $\{\sigma(\alpha) | \sigma \in Aut_R \Omega_R\}$ and let $f = \prod_{i=1}^n (x - \alpha_i)$. Note that f is an indecomposable polynomial with coefficients in R since $\{\alpha_1, \ldots, \alpha_n\}$ is a transitive $Aut_R \Omega_R$-set. By Theorem 3.6 on page 107 of [DI] the rank of S over R is n. Let $\varphi : R[x]/(f) \to R[\alpha]$ be the surjective R- algebra homomorphism given by $g \mapsto g(\alpha)$. Since $R[\alpha]$ is projective, $R[x]/(f) \simeq ker\varphi \oplus R[\alpha]$. The R-algebra $R[x]/(f)$ is free of rank n. Thus $ker\varphi = \{0\}$. Hence $f \in R[x]$ is an indecomposable separable polynomial, $f(\alpha) = 0$, and $R[x]/(f) \simeq S$. This completes the proof.

One consequence of Lemma 1.2 on page 22 of [HM] is that if a separable polynomial factors into two indecomposable separable polynomials then these factors are distinct and each of these factors is separable. We now characterize those rings over which the product of two distinct indecomposable separable polynomials is separable.

THEOREM 3.4   The product of any two distinct indecomposable separable polynomials in S is separable if and only if S is a field.

Proof: First assume that the product of any two distinct indecomposable separable polynomials in S is separable. Let a be a nonzero element in S. The polynomials x and $x - a$

are indecomposable and separable. Thus $x(x - a)$ is separable. So by Lemma 1.2 on page 22 of [HM] there exist $u, v \in S[x]$ such that $u \cdot (x - a) + v \cdot (x) = 1$. Hence $u(0) \cdot (-a) = 1$ and S is a field.

Conversely assume that S is a field. Let f and g be distinct monic indecomposable polynomials in $S[x]$. Since S is a field f and g are irreducible in S[x]. Thus there exist $u, v \in S[x]$ such that $uf + vg = 1$ since $S[x]$ is a Euclidean domain. If we further assume that f and g are separable then fg is separable by the proof of Lemma 1.5 on page 23 of [HM].

We now give a practical method for computing inertial degrees.

THEOREM 3.5   If f is an indecomposable separable polynomial in $R[x]$, M is a maximal ideal of R, and $f_1, \ldots, f_n$ are monic polynomials in $R[x]$ such that $\overline{f} = \overline{f}_1 \cdot \ldots \cdot \overline{f}_n$ is the unique factorization of $\overline{f}$ in R/M, and $S = R[x]/(f)$ then:

(i) The maximal ideals in S which lie over M are precisely the ideals of the form $M_i = M \cdot S + (f_i + (f)) \cdot S$;

(ii)$f(M_i|M) = deg f_i$.

Proof: For $i \in \{1, \ldots, n\}$ let $\pi_i : R[x]/(f) \to (R/M)[x]/(\overline{f}_i)$ be the obvious surjection and let $I_i = ker\pi_i$. Note that

$$S/(M \cdot S) \simeq (R/M)[x]/(\overline{f}_1) \times \ldots \times (R/M)[x]/(\overline{f}_n).$$

By Theorem 2.1 $M \cdot S$ is the intersection of all maximal ideals of S which lie over M. Thus $I_1, \ldots, I_n$ are precisely the the maximal ideals of S which lie over M. It is clear that each I_i contains M_i. If $g \in ker\pi_i$ then there exists $k \in R[x]$ such that $f_i \cdot k - g$ has coefficients in M. So $g \in M_i$ and $I_i = M_i$. This proves (i). Since $S/M_i \simeq (R/M)[x]/(\overline{f}_i)$ we have proven (ii).

We end this section with two applications of Theorem 3.5. First let $\mathbf{Z}$ denote the ring of integers, $T = \mathbf{Z}/9\mathbf{Z}$, M' denote the unique maximal ideal of T, and $f = x^2 + x + 2$. We have already noted that f is a separable indecomposable in $T[x]$ and $\overline{f}$ is irreducible in $(T/M')[x]$. So by Theorem 3.5, the ring $T[x]/(f)$ has one maximal ideal which lies over M' and this maximal ideal has inertial degree two over M'.

Next let T be the localization of the integers at the prime ideal which is generated by 2, M' denote the unique maximal ideal of T, and $f, u, v \in T[x]$ be given by $f = x^2 + x + 2, u = 4/7$, and $v = -2/7x - 1/7$. Note that f is separable since $uf + vf' = 1$. Clearly f is indecomposable in $T[x]$. Note that $\overline{f} = (x) \cdot (x + 1)$ is the unique factorization of $\overline{f}$ in $(T/M')[x]$. So by Theorem 3.5 the ring $T[x]/(f)$ has two maximal ideals which lie over M'. Both of these maximal ideals have inertial degree one.

4: Weak Henselization

If S is a connected ring then we let Ω_S denote a separable closure of S. Throughout this section we assume that R is a (not necessarily Noetherian) local ring with maximal ideal M.

DEFINITION 4.1   If M' is a maximal ideal of $\Omega_R, T = \Omega_R^{D(M')}$, and $M'' = M' \cap T$ then we call the ring $T_{M''}$ a weak Henselization of R.

In this section we will study the weak Henselizations of R. Let $R^{\#}$ denote a weak Henselization of R. We begin this section by considering some of the properties of local rings which are equal to their weak Henselizations. We pay particular attention to the separable polynomials over such rings. In the second part of this section we consider some of the properties of the R-algebra $R^{\#}$. In particular, we will show that a weak Henselization of R is a canonical object.

THEOREM 4.2   A weak Henselization of R is a local ring with residue class field isomorphic to the residue class field of R.

Proof: Let M' be a maximal ideal of $\Omega_R, T = \Omega_R^{D(M')}$, and $M'' = M' \cap T$. It is clear that $T_{M''}$ is a local ring. By Theorem 2.8 $T/M'' \simeq R/M$. Hence $T_{M''}/(M'' T_{M''}) \simeq R/M$.

THEOREM 4.3 $R = R^{\#}$ if and only if Ω_R is local.

Proof: This follows immediately from the definition of a weak Henselization.

We now consider an example. Let S be a finite local ring and let T be a finite projective connected separable extension of S. By Theorem vii on page 111 in [Mc], T is a local ring. So by Theorem 4.3, $S = S^{\#}$.

Next we turn our attention to the separable polynomials over local rings which are equal to their weak Henselizations. By the preceding theorem this is equivalent to studying the separable polynomials over local rings which have a local separable closure. If f is a separable polynomial over R and if $\overline{f}$ is indecomposable over R/M it is clear that f is indecomposable over R. If R has a local separable closure the converse is also true.

THEOREM 4.4   If Ω_R is local and if f is a separable polynomial with coefficients in R then:

(i) If f is indecomposable in $R[x]$ then $\overline{f}$ is indecomposable in $(R/M)[x]$;

(ii) If $f_1 \cdot \ldots \cdot f_n$ is the unique decomposition of f in $R[x]$ then $\overline{f}_1 \cdot \ldots \cdot \overline{f}_n$ is the unique decomposition of $\overline{f}$ in $(R/M)[x]$.

Proof: (i) Assume $\overline{f}_1 \cdot \ldots \cdot \overline{f}_n$ is the unique decomposition of f in $(R/M)[x]$. By Lemma 1.2 on page 22 of [HM] the f_i are pairwise relatively prime. So by the Chinese Remainder Theorem, $(R/M)[x]/(\overline{f}) \simeq (R/M)[x]/(\overline{f}_1) \times \ldots \times (R/M)[x]/(\overline{f}_n)$. By assumption, $R[x]/(f)$ is local. Hence $n = 1$.

(ii) This follows immediately from part (i).

We say that R satisfies the primitive element condition if every free, finite, separable, connected extension S of R has a primitive element. Note that by Theorem 3.3 this is equivalent to the condition that for every such S there exists an indecomposable separable polynomial $f \in R[x]$ such that $S \simeq R[x]/(f)$. On page 170 of [W] there is an example of a local ring which does not satisfy the primitive element condition.

THEOREM 4.5   If R/M is an infinite field and S is a finite, projective, separable extension of R then there exists a separable polynomial $f \in R[x]$ such that $S \simeq R[x]/(f)$.

Proof: This is Lemma 3.1 on page 471 of [J].

COROLLARY 4.6   If R/M is an infinite field then R satisfies the primitive element condition.

THEOREM 4.7   If Ω_R is local then R satisfies the primitive element condition.

Proof: Let S be a free, finite, separable, connected extension of R. Then $S/(MS)$ is a separable field extension of R/M. Hence there exists $\alpha \in S$ such that $S/(MS) = (R/M)[\alpha + MS]$. By Nakayama's Lemma, $S = R[\alpha]$.

LEMMA 4.8   If Ω_R is a local ring with maximal ideal M' then Ω_R/M' is a separable closure of R/M.

Proof: First we show that Ω_R/M' is a locally separable extension of R/M. Let $\alpha_1 + M', \ldots, \alpha_n + M' \in \Omega_R/M'$. There exists a finite, projective, separable, connected extension T of R such that $\alpha_1, \ldots, \alpha_n \in T$. By Theorem 7.1 on page 72 in [DI] $T/(MT)$ is a finite projective separable extension of R/M. Note that we can embed $T/(MT)$ in $\Omega_R/(M \cdot \Omega_R)$. By Theorem 2.1, $\Omega_R/(M \cdot \Omega_R)$ is isomorphic to Ω_R/M'. Thus Ω_R/M' is a locally separable extension of R/M.

Now we show that Ω_R/M' is separably closed. Assume by way of contradiction that there exists a finite projective connected separable extension T of Ω_R/M' such that the rank of T over Ω_R/M' is greater than one. By Corollary 2.4 on page 49 of [DI], T is a separable field extension of Ω_R/M'. Hence there exists a monic separable irreducible polynomial f with coefficients in Ω_R/M' such that T is isomorphic to $(\Omega_R/M')[x]/(f)$. Note that the degree of f must be greater than one. Let g be a monic polynomial with coefficients in Ω_R such that $g = f$ modulo M'. Clearly g is indecomposable over Ω_R. Further, by Theorem 2.2 on page 467 of [J], g is separable. Since Ω_R is separably closed the degree of g is one. Thus the degree of f is one. This is a contradiction. Hence Ω_R/M' is separably closed. This completes the proof.

Next we determine the separable closure of a homomorphic image of a ring which has a local separable closure.

THEOREM 4.9   If Ω_R is a local ring and I is an ideal of R then $\Omega_R/(I\Omega_R)$ is a separable closure of R/I.

Proof: An argument similar to the one given in the proof of Lemma 4.8 shows that that $\Omega_R/(I\Omega_R)$ is a locally separable extension of R/I. To see that $\Omega_R/(I\Omega_R)$ is separably closed let $S = \Omega_R/(I\Omega_R)$ and let Q be the unique maximal ideal of S. Assume by way of contradiction that there exists a finite projective connected separable extension T of S such that the rank of T over S is greater than one. By Lemma 4.8 S/Q is a separable closure of R/M. Hence S/Q must be an infinite field. So by Corollary 4.6 there exists a polynomial f with coefficients in S such that T is isomorphic to $S[x]/(f)$. As in the proof of Lemma 4.8 one can show that the degree of f is one. This contradiction shows that S is separably closed. This completes the proof.

COROLLARY 4.10   If $R = R^\#$ and S is a homomorphic image of R then $S = S^\#$.

We now show that if Ω_R is local then the absolute Galois group of R is isomorphic to the absolute Galois group of R/M.

THEOREM 4.11   If Ω_R is a local ring then $Aut_R\Omega_R \simeq Aut_{R/M}\Omega_{R/M}$.

Proof: Since Ω_R is local $D(M') = Aut_R\Omega_R$. So by Theorem 2.7

$$Aut_R\Omega_R \simeq Aut_{R/M}\Omega_R/M'.$$

By Lemma 2.8 Ω_R/M' is a separable closure of R/M. Hence

$$Aut_{R/M}\Omega_{R/M} \simeq Aut_{R/M}\Omega_R/M'.$$

This completes the proof.

In the next three theorems we consider local rings which have both a finite residue class field and a local separable closure.

THEOREM 4.12   Assume that Ω_R is local and R/M is a finite field. Then for every natural number n there exists a unique Galois extension T of rank n over R. Furthermore, $Aut_R T$ is a cyclic group of order n.

Proof: Recall that if F is a finite field then for every natural number n there exists a unique Galois extension L of rank n over F. Furthermore, $Aut_F L$ is a cyclic group of order n. The theorem now follows from Theorem 3.6 on page 107 of [DI] and Theorem 4.11.

THEOREM 4.13   If Ω_R is local, R/M is a finite field, and S is a finite projective connected extension of R then S is separable if and only if S is Galois.

Proof: If S is Galois then S is separable by definition. Conversely assume that S is separable. We may embed S in a finite Galois extension T of R such that T in contained in Ω_R. By the last theorem $Aut_R T$ is cyclic. Hence $Aut_S T$ is normal in $Aut_R T$ and S is a normal extension of R.

We now show that if R/M is a finite field then either R is equal to its weak Henselization or else Ω_R has infinitely maximal ideals.

THEOREM 4.14   If R/M is a finite field and Ω_R has finitely many maximal ideals then Ω_R is local.

Proof: Let $M_1, \ldots, M_g$ be the distinct maximal ideals of Ω_R. For $i \in \{1, \ldots, g\}$ let $m_i \in M_i - \cup_{j \neq i} M_j$. We know that there exists a finite Galois extension S of R such that $m_1, \ldots, m_g \in S$. For $i \in \{1, \ldots, g\}$ let $M_i' = M_i \cap S$. Note that $M_1', \ldots, M_g'$ are the distinct maximal ideals of S. Let $n = rank_R S$ and let p be a prime number which does not divide n. Let $\overline{h}$ be an irreducible separable polynomial over R/M of degree p. Note that h is an indecomposable separable polynomial over R of degree p. By our choice of p we see that p and $f(M_1')$ are relatively prime. Since R/M is a finite field, this implies that h is irreducible over S/M_1'. Thus h is indecomposable over S. We may embed $S[x]/(h)$ in a finite connected Galois extension T of S. Let $M_1'' = M_1 \cap T$. Note that both S and T have exactly g maximal ideals. Hence $Aut_S T = D(M_1''|M_1') \simeq Aut_{S/M_1} T/M_1''$. Thus the Galois group of T over S is cyclic and $S[x]/(h)$ is a normal extension of S. Hence $p = rank_S S[x]/(h) = (f(M_1''|M_1')) \cdot g$ and since p is prime it follows that $g = 1$. This completes the proof.

Next we turn our attention to local rings with infinite residue class field.

THEOREM 4.15   If R/M is an infinite field then the following conditions are equivalent:

(i) Ω_R is local;

(ii) $R = R^{\#}$;

(iii) If f is a separable polynomial in $R[x]$ and if there exist monic polynomials g_0, h_0 in $(R/M)[x]$ such that $\overline{f} = g_0 \cdot h_0$ then there exist monic polynomials g, h in $R[x]$ such that $f = g \cdot h, \overline{f} = f_0$, and $\overline{g} = g_0$;

(iv) If f is an indecomposable separable polynomial in $R[x]$ then $\overline{f}$ is an irreducible polynomial in $(R/M)[x]$;

(v) If f is a separable polynomial in $R[x]$ and $f_1 \cdot \ldots \cdot f_n$ is the unique decomposition of f in $R[x]$ then $\overline{f}_1 \cdot \ldots \cdot \overline{f}_n$ is the unique factorization of $\overline{f}$ in $(R/M)[x]$;

(vi) Every finite, projective, separable extension of R is a direct sum of finite, projective, separable, local rings;

(vii) If f is a separable polynomial in $R[x]$ and β is an element in R/M such that $\overline{f}(\beta) = 0$ then there exists an element α in R such that $f(\alpha) = 0$ and $\overline{\alpha} = \beta$.

(viii) If g is a separable polynomial in $R[x]$ such that $\overline{g}(0) = 0$ then there exists an element γ in R such that $g(\gamma) = 0$ and $\overline{\gamma} = 0$.

Proof: $(i) \Rightarrow (ii)$ This follows from Theorem 4.3.

$(ii) \Rightarrow (iii)$ Let $f_1 \cdot \ldots \cdot f_n$ be the unique decomposition of f in $R[x]$. By Theorem 4.4, $\overline{f}_1 \cdot \ldots \cdot \overline{f}_n$ is the unique decomposition of $\overline{f}$ in $(R/M)[x]$. We may reindex if necessary to obtain $\overline{f}_1 \cdot \ldots \cdot \overline{f}_i = g_0$ and $\overline{f}_{i+1} \cdot \ldots \cdot \overline{f}_n = h_0$ for some $i \in \{1, \ldots, n\}$.

$(iii) \Rightarrow (iv)$ and $(iv) \Rightarrow (v)$ These implications are clear.

$(v) \Rightarrow (vi)$ Let T be a finite, projective, separable extension of R. By Theorem 4.5 there exists a separable polynomial $f \in R[x]$ such that $R[x]/(f) \simeq T$. Write $f = f_1 \cdot \ldots \cdot f_n$ for the unique decomposition of f in $R[x]$. By Lemma 1.2 on page 22 of [HM] and the Chinese Remainder Theorem we have

$$R[x]/(f) \simeq R[x]/(f_1) \times \ldots \times R[x]/(f_n).$$

By assumption each $\overline{f}_i$ is irreducible over R/M. Hence by Theorem 3.5, each $R[x]/(f_i)$ is local.

$(vi) \Rightarrow (vii)$ Let $f_1, \ldots, f_n$ be monic polynomials in $R[x]$ such that $f_1 \cdot \ldots \cdot f_n$ is the unique decomposition of f in $R[x]$, let $\beta \in R/M$ such that $f(\beta) = 0$, and let $i \in \{1, \ldots, n\}$ such that $\overline{f}_i(\beta) = 0$. Assume by way of contradiction that the degree of f_i is greater than one. Then by Theorem 3.5, the finite, projective, connected, separable extension $R[x]/(f_i)$ is not a local ring. This contradicts the assumption that every finite, projective, separable extension of R is a direct sum of finite, projective, separable, local rings. Thus the degree of f_i is equal to one. So there exists $\alpha \in R$ such that $f_i = x - \alpha$. Note that $f(\alpha) = 0$ and $\overline{\alpha} = \beta$.

$(vii) \Rightarrow (viii)$ This implication is clear.

$(viii) \Rightarrow (i)$ Let S be a finite connected Galois extension of R of rank greater than one and let M'' be a maximal ideal of S. By Corollary 4.6 there exists an indecomposable separable polynomial $g \in R[x]$ such that $S^{D(M'')} \simeq R[x]/(g)$. Furthermore, by Theorem 2.8 $f(M'' \cap S^{D(M'')}) = 1$. Thus by Theorem 3.5, there exists $\beta \in (R/M)$ such that $\overline{g}(\beta) = 0$.

Let $\alpha \in R$ such that $\overline{\alpha} = \beta$. By Theorem 1.1 on page 22 of [HM] there exist $u, v \in R[x]$ such that $ug + vg' = 1$. Note that $u(x + \alpha)g(x + \alpha) + v(x + \alpha)g'(x + \alpha) = 1$. Thus by

Theorem 1.1 on page 22 of [HM] $g(x + \alpha)$ is separable. Now $\overline{g}(\overline{0 + \alpha}) = 0$ and so by assumption there exists $\gamma \in R$ such that $g(\gamma + \alpha) = 0$ and $\overline{\gamma} = 0$. Thus $x - (\gamma + \alpha)$ is a linear factor of g. Since g is indecomposable, we conclude that $g = x - (\gamma + \alpha)$. Thus $[S^{D(M'')} : R] = 1$. Hence by part (1) of Theorem 2.6, S is a local ring. Since S was chosen to be an arbitrary finite projective Galois extension of R we conclude that Ω_R is local. This completes the proof.

We now turn our attention to the R-algebra $R^\#$. By Proposition 2.1 on page 40 in [DI] there exists $\sum_{i=1}^j e_i \otimes e'_i \in T \otimes_R T$ such that $\sum_{i=1}^j e_i \otimes e'_i$ is a separability idempotent for T. Note that $\sum_{i=1}^j \frac{e_i}{1} \otimes \frac{e'_i}{1}$ is a separability idempotent for $R^\#$. Thus we have the following result.

THEOREM 4.16   The ring $R^\#$ is a separable extension of R.

We now show that a weak Henselization of R is a canonical object.

THEOREM 4.17   If A and B are weak Henselizations of R then there exists a unique R-algebra isomorphism from A to B.

Proof: We need to construct two weak Henselizations of R. First let M' is a maximal ideal of $\Omega_R, T = \Omega_R^{D(M')}$, $M'' = M' \cap T$, and $A = T_{M''}$. Next let Ω'_R be a separable closure of R, Q' be a maximal ideal of Ω'_R, $S = \Omega_R'^{D(Q')}$, $Q'' = Q' \cap S$, and $B = S_{Q''}$. We know that there exists an R-algebra isomorphism σ from Ω_R to Ω'_R. By Lemma 2.2 there exists $\tau \in Aut_R\Omega'_R$ such that $\tau(\sigma(M')) = Q'$. Note that the map $\varphi : A \to B$ given by $\varphi(\alpha/\beta) = \tau(\sigma(\alpha))/\tau(\sigma(\beta))$ is an R-algebra isomorphism.

To see that φ is unique it is enough to show that $Aut_RA = \{1\}$. Given $\gamma \in Aut_RA$ let $\overline{\gamma} \in Aut_{R/M} A/M''$ by defined by $\overline{\gamma}(\rho + M'') = \gamma(\rho) + M''$. By Theorem 4.2 $A/M'' \simeq R/M$. Hence the map $\gamma \mapsto \overline{\gamma}$ is a group homomorphism onto the trivial group. Furthermore by Theorem 4.16 A is a separable R-algebra and thus by Theorem 2.5 the kernel of this homomorphism is trivial. Hence $Aut_RA = \{1\}$. This completes the proof.

THEOREM 4.18   A weak Henselization of R is a flat R-module.

Proof: Let M' be a maximal ideal of $\Omega_R, T = \Omega_R^{D(M')}$, and $M'' = M' \cap T$. Let $\alpha_1/\beta_1, \ldots, \alpha_n/\beta_n \in T_{M''}$. There exists a free separable extension S of R such that S is contained in T and $\alpha_1, \ldots \alpha_n, \beta_1, \ldots \beta_n \in S$. Since S is free $\otimes_R S$ is exact. Clearly $\otimes_S S_Q$ is exact where $Q = M' \cap S$. Hence $\otimes_R S_Q$ is exact. Note that $\alpha_1/\beta_1, \ldots, \alpha_n/\beta_n \in S_Q$. This shows that for any finite number of elements in $T_{M''}$ there exists an R-submodule A of $T_{M''}$ which contains these elements and such that $\otimes_R A$ is exact. This implies that $\otimes_R T_{M''}$ is exact.

5: Henselization

Throughout this section we will assume that R is a local domain which is integrally closed in its quotient field and that M is the maximal ideal of R. In this section we will study the relationship between the weak Henselizations of R and the Heselizations of R.

For the remainder of this section, let Λ_R denote the set of elements in a separable closure of the quotient field of R which are integral over R, let M' be a maximal ideal of Λ_R, and let $M'' = M' \cap \Lambda_R^{D(M')}$. The ring $(\Lambda_R^{D(M')})_{M''}$ is called a Henselization of R. We let R^* denote a Henselization of R.

If P is a maximal ideal of Λ_R we write $\Gamma(P)$ for the intersection of all closed normal subgroups of $Aut_R\Lambda_R$ which contain $E(P)$.

LEMMA 5.1   The ring $\Lambda_R^{\Gamma(M')}$ is a separable closure of R.

Proof: Let $\Gamma = \Gamma(M')$ and let F denote the quotient field of R. First we show that Λ_R^Γ is a locally separable extension of R. Let $\alpha_1, \ldots, \alpha_n$ be elements in Λ_R^Γ, $K \subseteq \Omega_F^\Gamma$ be a finite Galois extension of F such that $\{\alpha_1, \ldots, \alpha_n\} \in K$, and $T = \Lambda_R^\Gamma \cap K$. Note that $Aut_R T$ is finite since $Aut_F K$ is finite and that T is a normal extension of R since K is a normal extension of F. Let $P = M' \cap T$ and let $\sigma \in E(P)$. By the remark on page 339 of [B] there exists $\hat{\sigma} \in E(M')$ such that $\hat{\sigma}|_T = \sigma$. Note that $\hat{\sigma}(\beta) = \beta, \forall \beta \in \Lambda_R^\Gamma$. Hence $\sigma = 1$. Now if P' is another maximal ideal of T then there exists a $\tau \in Aut_R T$ such that $\tau E(P')\tau^{-1} = E(P)$. Thus $E(P') = \{1\}$. So by Lemma 2.3 T is separable. Thus by Corollary 4.2 on page 473 in [J] T is projective. Hence Λ^Γ is locally separable.

To see that Λ_R^Γ is a separable closure of R let T be a finite connected Galois extension of Λ_R^Γ. By Corollary 4.2 on page 473 in [J] T is a domain. Furthermore, by Proposition 1.1 on page 40 in [DI] there exists $\sum_{i=1}^j e_i \otimes e_i' \in T \bigotimes_R T$ such that $\sum_{i=1}^j e_i \otimes e_i'$ is a separability idempotent for T. Note that $\sum_{i=1}^j \frac{e_i}{1} \otimes \frac{e_i'}{1}$ is a separability idempotent for the quotient field of T. Thus by Proposition 1.1 on page 40 in [DI], the quotient field of T is a separable extension of F. Hence $T \subseteq \Lambda_R$

Note that if P is a maximal ideal of T then the inertial subgroup of P with respect to $Aut_R T$ is trivial. Thus $\Gamma \subseteq Aut_T\Lambda_R$ and $T \subseteq \Lambda_R^\Gamma$. Hence $T = \Lambda_R^\Gamma$. This completes the proof.

We now prove that any weak Henselization of R can be embeded in any Henselization of R.

THEOREM 5.2   There exists an R-algebra monomorphism from $R^\#$ to R^*.

Proof: This follows from Lemma 5.1 and the definition of a weak Henselization.

COROLLARY 5.3   If $R = R^*$ then $R = R^\#$.

We now give a condition which is equivalent to the condition that $R^\#$ is isomorphic to R^*.

THEOREM 5.4   The following are equivalent:

(i) There exists a unique R-algebra isomorphism from $R^\#$ to R^*;

(ii) $D(M') = \Gamma(M')$.

Proof: Let $\Gamma = \Gamma(M'), Q' = M' \cap \Lambda_R^\Gamma$, and $Q'' = M'' \cap (\Lambda_R^\Gamma)^{D(Q')}$.

$(i) \Rightarrow (ii)$ By Theorem 4.17 and the fact Henselizations are cannonical objects we have

$$((\Lambda_R^\Gamma)^{D(Q')})_{Q''} = ((\Lambda_R)^{D(M')})_{M''}.$$

Thus

$$(\Lambda_R^\Gamma)^{D(Q')} = (\Lambda_R)^{D(M')}$$

and

$$\Gamma = D(M').$$

$(ii) \Rightarrow (i)$ Note that

$$((\Lambda_R^\Gamma)^{D(Q')})_{Q''} = ((\Lambda_R^{D(M')})^{D(Q')})_{Q''}$$
$$= (\Lambda_R^{D(M')})_{M''}.$$

The implication now follows from Theorems 4.17 and the fact that Henselizations are cannonical objects.

COROLLARY 5.5   If R/M is infinite and $D(M') = \Gamma(M')$ then the following conditions are equivalent:

(i) If $f \in R[x]$ is monic and there exist monic polynomials $g_0, h_0 \in R/M[x]$ such that g_0 and h_0 are relatively prime and $\overline{f} = g_0 h_0$ then there exist monic polynomials $g, h \in R[x]$ such that $\overline{g} = g_0, \overline{h} = h_0$, and $f = gh$;

(ii) If $f \in R[x]$ is a separable polynomial and there exist monic polynomials $g_0, h_0 \in (R/M)[x]$ such that $\overline{f} = g_0 h_0$ then there exist monic polynomials $g, h \in R[x]$ such that $\overline{g} = g_0, \overline{h} = h_0$, and $f = gh$.

We conclude this section with an example. Assume that R is the ring of integers localized at the prime ideal which is generated by 3. On page 103 of [M] it is proven that there exists a subring S of Λ_R such that S is a finite normal extension of R and if Q is a maximal ideal of S then $D(Q)$ is not a normal subgroup of $Aut_R S$. Thus $D(M')$ is not a normal subgroup of $Aut_R \Lambda_R$. It follows that $D(M') \neq \Gamma(M')$. So by Theorem 5.2, the Henselization of R is not isomorphic to the weak Henselization of R.

References

[B] N. Bourbaki, *Commutative algebra*, Addison-Wesley, Reading, Mass., 1972.

[DI] F. Demeyer and E. Ingrahm, *Separable algebras over commutative rings*, Lecture Notes in Math. vol. 181, Springer-Verlag, New York, 1971.

[HM] D. K. Harrison and T. McKenzie, *Toward an arithmetic of polynomials*, Aequationes Math. **43** (1992), 21-37.

[J] G. Janusz, *Separable algebras over commutative rings*, Trans. Amer. Math. Soc. **122** (1966), 461-479.

[M] D. Marcus, *Number fields*, Springer-Verlag, New York, 1977.

[Mc] B. McDonald, *Finite rings with identity*, Marcel Dekker, New York, 1974

[W] E. Weiss, *Algebraic number theory*, McGraw-Hill, New York, 1962.

Faithful Representations of Lie Algebras over Power Series

GRAYDON NELSON   University of Oklahoma, Norman, Oklahoma

INTRODUCTION

This is an exposition of the authors doctoral thesis completed at the University of Oklahoma under the direction of Professor Andy Magid. We begin with a brief discussion of the work done by Dixon, Du Sautoy, Mann and Segal (DDMS) in showing that powerful pro-p groups are almost linear over the p-adic integers $\mathbf{Z}_p$. That is, G has a subgroup of finite index which injects into $\mathbf{Gl}_d(\mathbf{Z}_p)$ [3, Cor. 4.19, p.73].

They show that a topologically finitely generated pro-p group is a $\mathbf{Z}_p$–module with additive structure $+_G$ and bracket structure $(,)_G$.

Definition 0.1. *A topological group is a set G together with two structures:*

(1) $(G, \cdot)$ *is a group;*

(2) $(G, \ )$ *is a topological space;*

such that the two structures are compatible. Namely, the multiplication map $G \times G \to G; (a, b) \mapsto ab$ and the inverse map $G \to G; a \mapsto a^{-1}$ are continuous where $G \times G$ has the product topology.

They then provide a characterization of a pro-p group which gives some insight into the structure [3, Prop. 1.12, p.27].

Proposition 0.2. *A topological group G is a pro-p group if G is topologically isomorphic to an inverse limit of finite p-groups (i.e. $G = \varprojlim N_i$)*

To get an axiomatic characterization of linearity, DDMS define the following congruence system which models the principle congruence subgroups of $\mathbf{Gl}_n(\mathbf{Z}_p)$ [3, Defn. 1.15, p.29].

Definition 0.3. *Let G be a pro-p group. The lower p-series of G is given by $P_1(G) = G$ and for $i \geq 1$*

$$P_{i+1}(G) = \overline{P_i(G)^p [P_i(G), G]}.$$

There are two key properties which make pro-p groups linear. The first is a continuous homomorphism from $\mathbf{Z}_p$ into G given by $\lambda \mapsto g^\lambda$ $(g^\lambda = \lim_{n \to \infty} g^{a_n}$ where (a_n) is a sequence

181

of integers such that $\lim_{n\to\infty} a_n = \lambda$) [3, Prop.1.26, p.35]. The second is a *powerful condition* given by $[G, G] \leq \overline{G^p}$. [3, Defn. 3.1, p.52].

If $G = \overline{\langle a_1, \ldots, a_d \rangle}$ is a pro-p group with the powerful condition, then the derived subgroup is absorbed in the lower p- series and we have

$$P_i(G) = G^{p^{i-1}} = \{x^{p^{i-1}} | x \in G\} = \overline{\langle a_1^{p^{i-1}}, \ldots, a_d^{p^{i-1}} \rangle}$$

[3, Thm. 3.6, p.53]. From this mapping we can then conclude that

(1) $$G = \overline{\langle a_1 \rangle} \cdots \overline{\langle a_d \rangle}.$$

That is, G is the product of its procyclic subgroups $\overline{\langle a_1 \rangle}, \ldots, \overline{\langle a_d \rangle}$ [3, Prop.3.7, p.54].

Under this condition DDMS show that for a uniformly powerful pro-p group G with topological generating set $\{a_1, \ldots, a_d\}$, i.e. a powerful group where $|P_i(G) : P_{i+1}(G)| = |G : P_2(G)|$ [3, Defn. 4.1, p.64] , the map

(2) $$(\lambda_1, \ldots, \lambda_d) \mapsto a_1^{\lambda_1} \cdots a_d^{\lambda_d}$$

is a homeomorphism from $\mathbf{Z}_p^d$ to G [3, Thm. 4.9, p.68]. An additive structure, denoted $+_G$ and a bracket structure $(,)_G$ is then defined on G using a p^n-th power mapping $x \mapsto x^{p^n}$ from G to $P_{n+1}(G)$ [3, Defn. 4.12, p.70]. With this and (1) they conclude that $(G, +)$ is a free $\mathbf{Z}_p$–module with basis $\{a_1, \ldots, a_d\}$ [3, Thm. 4.17, p.73].

They then show that $Aut(G)$ may be identified with a subgroup of $\mathbf{Gl}_d(\mathbf{Z}_p)$ [3, Cor. 4.18, p.73] and that there exist a uniform open normal subgroup H of G with finite index such that $Z(H) = \mathbf{Z}_p^e$ where $e \leq d$. Then there is a homomorphism

$$\theta : G \to Aut(G) \times G/H$$

given by $\theta(g) = (g^*, gH)$ where g^* denotes the automorphism of H induced by conjugation with g. Since $ker(\theta) = Z(H)$ and $Aut(G) \leq \mathbf{Gl}_d(\mathbf{Z}_p)$, we have that the sequence

$$1 \to \mathbf{Z}_p^e \to G \to \mathbf{Gl}_n(\mathbf{Z}_p) \times G/H$$

is exact [3, Cor. 4.19, p.73]. From this we can conclude that $\theta(H) \leq \mathbf{Gl}_d(\mathbf{Z}_p)$.

In part II DDMS show that if G is a pro-p group of finite rank, then G admits a faithful linear representation

$$\psi : G \to \mathbf{Gl}_d(\mathbf{Q}_p)$$

where $\mathbf{Q}_p$ is the quotient field of $\mathbf{Z}_p$ [3, Thm.8.2, p.170].

They start by defining a norm on the group algebra $A = \mathbf{Q}_p[G]$ [3, Defn. 8.12, p.164] compatible with the topology on G so as to make A into a normed $\mathbf{Q}_p$ algebra. If $\widehat{A}$ is the completed group algebra of G then they show that there exists an associative algebra $\widehat{A}_0$ of $\widehat{A}$ such that $\Lambda = \log G \subseteq \widehat{A}_0$ where the maps

$$\log(1+x) = \sum_{n=1}^{\infty}(-1)^{n+1}\frac{x^n}{n}$$

and

$$exp(x) = \sum_{n=0}^{\infty}\frac{x^n}{n!}$$

are defined on A_0 and satisty the usual identities for log and exp [3, Defn. 7.20, p.142] and [3, Defn. 7.23, p.143].

Since $\widehat{A}_0$ is an associative algebra it can be given the structure of a Lie algebra over $\mathbf{Q}_p$ with its Lie bracket operation defined by $(x,y) = xy - yx$.

They then give the following lemma [3, Lemma 8.14, p.165].

Lemma 0.4. *For all $g, h \in G$ and $\lambda \in \mathbf{Z}_p$, we have in Λ*

  (1) $\log g + \log h = \log(g +_G h)$;
  (2) $\lambda \log g = \log g^\lambda$; *and*
  (3) $(\log g, \log h) = \log(g, h)_G$.

It is immediate that $(\Lambda, +, (,))$ is a $\mathbf{Z}_p$–Lie subalgebra of the Lie algebra $(\widehat{A}, +, (,))$ [3, Cor. 8.15, p.167]. It is also clear from Lemma 4 that log defines a bijection from G onto Λ and from this they conclude that

$$\log : (G, +_G) \rightarrow (\Lambda, +)$$

is a $\mathbf{Z}_p$–module isomorphism with the property that

$$\log(g, h)_G = (\log g, \log h).$$

From this we get the following corollary [3, Cor. 8.16, p.167].

Corollary 0.5. *(i) $(G, +_G, (,)_G)$ is a $\mathbf{Z}_p$–Lie algebra of dimension d. (ii) The mapping $\log : ((G, +_G, (,)_G) \rightarrow (\Lambda, +, (,))$ given by $g \mapsto \log g$ is a $\mathbf{Z}_p$–Lie algebra isomorphism. (iii) $(\Lambda, +, (,))$ has dimension d.*

Let $\mathbf{Q}_p\Lambda$ denote the $\mathbf{Q}_p$–vector subspace of $\widehat{A}$ spanned by Λ. By Corollary 5 Λ has dimension d as a $\mathbf{Z}_p$–vector space. From this we conclude that $\mathbf{Q}_p\Lambda$ is a $\mathbf{Q}_p$–vector space of dimension d. We need the following theorem by Ado [9, Ado's Thm., p.202].

Theorem 0.6. *Let L be a finite dimensional Lie algebra over a field k of characteristic zero. Then L admits a faithful finite dimensional linear representation*

$$\phi : L \to \mathbf{gl}_n(k)$$

where $\mathbf{gl}_n(k)$ is the algebra of all $n \times n$ matrices over k considered as a Lie algebra with the Lie bracket operation given by $(x, y) = xy - yx$, and ϕ is a Lie algebra homomorphism.

From Theorem 0.6 we have a faithful finite dimensional linear representation

$$\phi : \mathbf{Q}_p \Lambda \to \mathbf{gl}_d(\mathbf{Q}_p).$$

Since $G = \Lambda \hookrightarrow \mathbf{Q}_p \Lambda$ we have a faithful finite dimensional linear representation of G into $\mathbf{gl}_d(\mathbf{Q}_p)$. Let $\Lambda_0 = \phi^{-1}(\phi \Lambda \cap \mathbf{gl}_d(p\mathbf{Z}_p))$. Then $\exp \Lambda_0 \subseteq G$. We need the following Lemma [3, Lemma 8.18, p.168].

Lemma 0.7. *There exist $m \geq 1$ with the property that $P_m(G) \subseteq \exp \Lambda_0$.*

They then define a map $\psi = \exp \circ \phi \circ \log : G_m \to \mathbf{gl}_d(\mathbf{Q}_p)$, where $G_m \subseteq \exp \Lambda_0$ and show that ψ is a faithful linear representation of G_m into $\mathbf{Gl}_d(\mathbf{Q}_p)$ [3, Thm. 8.19, p.169] which lifts to G [3, Thm 8.20, p.170].

SECTION I

The action of $\mathbf{C}[[t]]$ on L.

Work done by Magid and Lubotsky with prounipotent groups shows a resemblence to that of pro-p groups and a similar axiomatic description could lead to a characterization of linear prounipotent groups [11].

A model for a uniform pro-p group is

$$G = \mathbf{Gl}_n(\mathbf{Z}_p) \cap I + p\mathbf{M}_n(\mathbf{Z}_p)$$

and a model for a prounipotent group is similarly defined by

$$G = \mathbf{Sl}_n(\mathbf{C}[[t]]) \cap I + t\mathbf{M}_n(\mathbf{C}[[t]]).$$

The local ring $\mathbf{C}[[t]]$ is used as a generalization of the local ring $\mathbf{Z}_p$. We want to define a map from $\mathbf{C}[[t]]$ into a prounipotent group G similar to the map from $\mathbf{Z}_p$ into a pro-p group G given by

$$\lambda \mapsto g^\lambda$$

which would define an action of $\mathbf{C}[[t]]$ on the prounipotent group G. The obvious choice is to define

$$f \mapsto A^f = \exp(f \log A)$$

where

$$\log A = -\sum_{k=1}^{\infty} \frac{(I - A)^k}{k}$$

and

$$\exp A = \sum_{k=0}^{\infty} \frac{A^k}{n!} \; .$$

There are several difficulties in working with these functions and an additive structure on G would make them unnecesary.

There is a satisfactory Lie theory for prounipotent groups. The Lie algebras of prounipotent groups are the pronilpotent Lie algebras. Hence, if L is a suitable pronilpotent Lie algebra, that is L is a complex Lie algebra with lower central series C^i such that L/C^i is finite dimensional and $L = \varprojlim L/C^i$, then an appropriate axiomatic description for the characterization of linear pronilpotent Lie algebras is a good step toward the characterization of linear prounipotent groups. This exposition provides such a characterization for these pronilpotent Lie algebras. As in pro-p groups, it turns out that a condition analogous to the powerful condition for pro-p groups and a mapping from $\mathbf{C}[[t]]$ into L similar to the mapping from $\mathbf{Z}_p$ into the pro-p group are the key components for linearity.

We start with

Definition 1. *Let L be a Lie algebra over the complex numbers with lower central series $C^0 = L$, $C^{i+1} = [L, C^i]$. L is a pro(finite dimensional) nilpotent Lie algebra if $L = \varprojlim L/C^i$ where L/C^i is finite dimensional as a vector space over the complex numbers for all $i \geq 0$.*

Unless otherwise stated, pronilpotent will mean pro(finite dimensional) nilpotent. The following discussion gives an axiomatic characterization for linearity conditions on L which induce a finite dimensional $\mathbf{C}[[t]]$–module structure on L.

It is well known that there exists a one-to-one correspondence between a vector space L with a linear transformation $t : L \to L$ and the $\mathbf{C}[t]$–module L where the action is given by

$$\sum_{k=0}^{n} a_k t^k \cdot v = \sum_{k=0}^{n} a_k t^k(v)$$

We extend this action and make L a $\mathbf{C}[[t]]$–module by defining

$$\sum_{k=0}^{\infty} a_k t^k \cdot v = \sum_{k=0}^{\infty} a_k t^k(v)$$

Since

$$\lim_{n\to\infty} \sum_{k=0}^{n} a_k t^k(v) = \sum_{k=0}^{\infty} a_k t^k(v)$$

the sequence

$$\left\{ \sum_{k=0}^{n} a_k t^k(v) \right\}_{n=0}^{\infty}$$

has a limit if L is complete in some topology in which the sequence is Cauchy.

The following definition provides a "powerful like" condition on L which will provide a topology in which L is complete and the above sequence converges.

Definition 2. *A pronilpotent Lie algebra L is powerful if there exist a linear transformation $t : L \longrightarrow L$ such that:*

(1) $t[x, y] = [x, ty]$;

(2) $[L, L] \subseteq tL$;

(3) $\cap_{i=0}^{\infty} t^i L = 0$.

Then $\{t^i L\}_{i=0}^{\infty}$ is a decreasing sequence of ideals which form a neighborhood basis of zero in L and give the $t^i L$–adic topology on L. Furthermore

$$\sum_{k=0}^{m} a_k t^k(v) - \sum_{k=0}^{n} a_k t^{(}v) = \sum_{k=n+1}^{m} a_k t^k(v) \in t^j L$$

for $m, n \geq j$. Therefore $\sum_{k=0}^{\infty} a_k t^k(v)$ is a Cauchy sequence in the $t^i L$–adic topology. So if we can show L comlete in this topology, that is $L = \varprojlim L/t^i L$, then the desired map will be well defined.

By Definition 2. $C = [L, L] \subseteq tL$ which implies $C^k \subseteq t^k L$ and since $L = \varprojlim L/C^i$, if follows that $\cap_{i=0}^{\infty} C^i = 0$. Also, the decending chain of ideals $t^i L_{i=1}^{\infty}$ forms a "$t-$ $-congruence system$" in L analogous to the $p--congruence system$ of normal subgroups given by Dixon, Du Sautoy, Mann and Segal [3, Defn. 6.1, p.93]. That is

(1) L/tL has finite dimension;

(2) $tL/t^i L$ is finite dimensional and every element is annihilated by a power of t for all $i \geq 1$;

(3) $\cap_{i=1}^{\infty} t^i L = 0$.

We simplify the above notation for the following argument. Our model will be the finite dimensional vector space $V_i = L/C^i$ over the complex numbers, P_i^{i+1} the canonical epimorphism from L/C^{i+1} to L/C^i, $T^i = t^i$, $T_i = \bar{t}$ the canonical linear operator on L/C^i induced by t, and $V = L$.

Let $V_i, P_i^{i+1}{}_{i=0}^{\infty}$ be an inverse system of finite dimensional vector spaces over the complex numbers and onto linear transformations of the vector space V such that

(1) $T = \lim T_i$;

(2) $\cap_{i=1}^{\infty} T^i V = 0$;

(3) $T_i P_i^{i+1} = P_i^{i+1} T_{i+1}$.

That is, the following diagram commutes.

$$
\begin{array}{ccc}
V_{i+1} & \xrightarrow{P_i^{i+1}} & V_i \\
T_{i+1} \downarrow & & \downarrow T_i \\
V_{i+1} & \xrightarrow[P_i^{i+1}]{} & V_i
\end{array}
$$

Upon linking together n commutative diagrams as above we have $P_i^{i+1} T_{i+1}^n = T_i^n P_i^{i+1}$. By the Jordan–Chevalley decomposition theorem $T_i = D_i + N_i$ where D_i and N_i are uniquely determined polynomials in T_i, D_i is diagonalizable, N_i is nilpotent, and $D_i N_i = N_i D_i$.

Our goal is to show that $L = \lim L/t^i L$. Since $C^i \subseteq t^i L$, it is enough to show that $t^{n_i} \subseteq C^i$ for some n_i.

To this end we consider the primary decomposition of

$$
V_i = \bigoplus_{j=1}^{s_i} V_{i,\alpha_j}(D_i)
$$

from the diagonalizable linear operator D_i and translate this to product notation. That is

$$
V_i = \prod_{\alpha \in C} V_{i,\alpha}(D_i).
$$

Next we observe that $V_{i,\alpha}(D_i), P_i^{i+1}{}_{i=1}^{\infty}$ forms a surjective inverse system of vector spaces and linear transformations and from this conclude that

$$
V = \lim V_i = \lim \prod_{\alpha \in C} V_{i,\alpha}(D_i).
$$

From this we show that the inverse limit and the product commute, i.e.

$$
\lim \prod \alpha \in C V_{i,\alpha}(D_i) = \prod_{\alpha \in C} \lim V_{i,\alpha}(D_i)
$$

and with this result prove that $T = N$.

Now we return to our powerful pronilpotent Lie algebra L. Since C^i is t–invariant

$$\bar{t} : L/C^i \longrightarrow L/C^i$$

is a well defined linear transformation. We denote $\bar{t} = t_i$ and

$$T = \varprojlim t_i : \varprojlim L/C^i \longrightarrow \varprojlim L/C^i.$$

If

$$\phi : L \longrightarrow \varprojlim L/C^i$$

is an isomorphism, then we have the following commutative diagram.

$$
\begin{array}{ccc}
L & \overset{\phi}{\longrightarrow} & \varprojlim L/C^i \\
{\scriptstyle t}\downarrow & & \downarrow{\scriptstyle T} \\
L & \overset{\phi}{\longrightarrow} & \varprojlim L/C^i \\
{\scriptstyle \pi_j}\downarrow & & \downarrow{\scriptstyle P_j} \\
L/C^j & \overset{i}{\longrightarrow} & L/C^j
\end{array}
$$

That is $P_j\phi = \pi_j$ and $\phi t = T\phi$. Now

$$T^n = (\varprojlim t^i)^n = \varprojlim t^n$$

and $\phi t^n = T^n \phi$ so it follows that

$$t^{n_j}(v) + C^j = \pi(t^{n_j}(v)) + C^j = P_j\phi t^{n_j}(v)$$

$$= P_j T^{n_j}((v + C^i)_{i=1}^\infty) = P_j((t^{n_j}(v) + C^i)_{i=1}^\infty)$$

$$= t^{n_j}(v) + C^j = N^{n_j}(v) + C^j = C^j$$

which implies that

$$t^{n_j}(v) + C^j = C^j.$$

That is

$$t^{n_j} L \subseteq C^j.$$

Therefore L is complete in the $t^i L$–adic topology and the action of $\mathbf{C}[[t]]$ on L is well define.

SECTION II
The $\mathbf{C}[[t]]$–module structure of L.

To show that the action of $\mathbf{C}[[t]]$ on L induces a module structure, the following technical proposition is used to prove a key Lemma.

Proposition 1. *Let L be a powerful pronilpotent Lie algebra. Then*

(1) $t(\sum_{i=0}^{\infty} a_i t^i(v)) = \sum_{i=0}^{\infty} a_i t^{i+1}(v)$;

(2) *If $b \in \mathbf{C}$, then* $b\sum_{i=0}^{\infty} a_i t^i(v) = \sum_{i=0}^{\infty} b a_i t^i(v)$;

(3) *If $x_n \mapsto x$, $y_n \mapsto y$ in the $t^i L$-adic topology, then $x_n - y_n \mapsto x - y$;*

(4) $\sum_{k=0}^{n}(\sum_{i+j=k} a_i b_j) t^k(v) = \sum_{i=0}^{n} \sum_{j=0}^{n-i} a_i b_j t^{i+j}(v)$.

In proving the $\mathbf{C}[[t]]$–module structure of L the difficult step is showing the action is associative (i.e. $(gh)v = g(hv)$ for $g, h \in \mathbf{C}[[t]]$, and $v \in L$). It turns out that the key is the following lemma.

Lemma 1. *If L is a powerful pronilpotent Lie algebra, then*

$$\sum_{k=0}^{n}(\sum_{i+j=k} a_i b_j) t^k(v) - \sum_{i=0}^{n} a_i t^i(\sum_{j=0}^{\infty} b_j t^j(v))$$

$$= t^{n+1}(\sum_{j=n+1}^{\infty} t^{j-(n+1)}(-a_0 b_j v) + \sum_{j=n}^{\infty} t^{j-n}(-a_1 b_j v) + \cdots + \sum_{j=1}^{\infty} t^{j-1}(-a_n b_j v)).$$

From Lemma 1 we have

Theorem 1. *Let L be a powerful pronilpotent Lie algebra. Then $\varprojlim L/t^i L$ is a $\mathbf{C}[[t]]$–module.*

To show that L is finitely generated we need the following theorem from *Commutative Ring Theory* by H. Matsumura. [12, Thm. 8.4, p.58]

Theorem 2. *Let A be a ring, I an ideal and M an A-module. Suppose that A is I-adically complete and M is separated for the I-adic topology. If M/IM is generated over A/I by $\overline{w}_1, \overline{w}_2, \ldots, \overline{w}_n$ and $w_i \in M$ is an arbitrary inverse image of $\overline{w}_i$ in M, then M is generated over A by $w_i, w_2, \ldots, w_n$.*

Theorem 3. *Let L be a powerful pronilpotent Lie algebra. Then L is finitely generated as a $\mathbf{C}[[t]]$-module.*

Proof. Let $A = \mathbf{C}[[t]]$, $I = (t)$, and $M = L$. It is clear that $t^i L \subseteq (t)^i L$, so we will only show that $(t)^i L \subseteq t^i L$. Given $m \in (t)^i L$ we have

$$y = \sum_{j=i}^{\infty} a_j t^j(m) = t^i \sum_{j=i}^{\infty} a_j t^{j-i}(m) \in t^i L.$$

Thus $(t)^i L = t^i L$ and by assumption $\cap_{i=0}^{\infty} t^i L = 0$, therefore L is separated in the $t^i L$-adic topology. Now $\mathbf{C}[[t]]/(t) = \mathbf{C}$ and by assumption L/C^i is finitely generated over $\mathbf{C}$. Applying Theorem 2 shows that L is finitely generated over $\mathbf{C}[[t]]$.

Since $\mathbf{C}[[t]]$ is a principal ideal domain where the only ideals are t^d for $d \geq 0$, by the primary decomposition theorem for modules over a principal ideal domain we have

$$L = \mathbf{C}[[t]]^n \times \mathbf{C}[t]/(t^{d_1}) \times \cdots \times \mathbf{C}[t]/(t^{d_r}).$$

SECTION III
Generating sets and the linear structure
of a powerful pronilpotent Lie algebra

We briefly turn away from the main goal of finding a faithful finite dimensional linear representation of L and discuss generating sets of powerful pronilpotent Lie Algebras.

In their theory of Analytic pro–p Groups, Dixon, Du Sautoy, Mann, and Segal show if G is a *profinite group* and $\Phi(G)$ is the *Frattini subgroup* of G, then X generates G topologically if and only if $X\Phi(G)/\Phi(G)$ generates $G/\Phi(G)$ topologically. Similarly for a powerful pronilpotent Lie algebra L we show that X generates L topologically as a Lie algebra over $\mathbf{C}$ if and only if the image of X in L/C generates L/C as a vector space over $\mathbf{C}$.

Let L be a finite dimensional nilpotent Lie algebra and S a maximal subalgebra of L. Since L is nilpotent, $Z_n(L) = L$ for some n. Also S maximal implies there exist i such that $Z_i(L) \subseteq S$ and $Z_{i+1}(L) \nsubseteq S$. Choose $y \in Z_{i+1}(L) - S$. Then

(*) $[L, y] \subseteq Z_i(L) \subseteq S.$

Let $S + \mathbf{C}y$ be the vector space spanned by S and y which is also closed under Lie product multiplication. Then $S + \mathbf{C}y$ is a subalgebra of L and $S \subseteq S + \mathbf{C}y$ which implies $S + \mathbf{C}y = L$.

Recall that L is generated by $X \subseteq L$ as a Lie albebra over $\mathbf{C}$ if ever element of L can be written as a linear combination of the Lie products from X.

Then using (*) and the above fact it can be shown that

(**) $[L, L] \subseteq S.$

The following discussion concerning Topological Groups is taken from [12, Section 8, p.55] and [5, Chapter II, p.28]. Let A be a ring, M an A–module, and $\{M_i\}_{i=1}^\infty$ a family of submodules which forms a fundamental system of neighborhoods of 0 and makes M into a topological group. This topology is called the M–adic of *linear topology* on M. If $N \subseteq M$ is a submodule, then the closure of N in M is given by

$$\overline{N} = \cap_{i=1}^\infty (M_i + N).$$

Also, the quotient topology on M/N is equal to the linear topology on M/N given by $\{(M_i + N)/N\}_{i=1}^{\infty}$.

We use the following notation. Let $X \subseteq L$. If X generates L as a Lie algebra, then $\langle\langle X \rangle\rangle = L$. If X generates L as a vector space, then $\langle X \rangle = L$. If X topologically generates L, then $\overline{\langle\langle X \rangle\rangle} = L$ or $\overline{\langle X \rangle} = L$.

We have L a topological group (in the C^i–adic topology) and

$$L - C = \cup\{x + C | x \in L - C\}$$

therefore C is closed as well as open in L. Thus L/C is discrete in the quotient topology which is the same as the linear topology given by $\{(C^i + C)/C\}_{i=1}^{\infty}$ and $\pi_1 : L \longrightarrow L/C$ is a continuous epimorphism of Lie algebras. Also $[x + C, y + C] = [x, y] + C$ implies L/C is an abelian Lie and therefore $\pi_1\langle\langle X \rangle\rangle = \langle \pi_1(X) \rangle$. Similarly L/C^i is a discrete topological group in the quotient topology and $\pi_1 : L \longrightarrow L/C^i$ is a continuous epimorphism. In addition $\pi_1^i : L/C^i \longrightarrow L/C$ is a continuous epimorphism such that the following diagram commutes.

$$
\begin{array}{ccc}
L & \xrightarrow{\pi_i} & L/C^i \\
{\scriptstyle id}\downarrow & & \downarrow{\scriptstyle \pi_1^i} \\
L & \xrightarrow{\pi_1} & L/C
\end{array}
$$

That is $\pi_1^i \pi_i = \pi_1$.

We now give the first main result of this section.

Theorem 4. *Let L be a pronilpotent Lie algebra over the complex numbers and $X \subseteq L$. Then X topologically generates L as a Lie algebra if and only if $\pi_1(X)$ generates L/C as a vector space over the complex numbers.*

Proof. ($\Rightarrow$) Let $y + C \in L/C$ where $y \in L = \overline{\langle\langle X \rangle\rangle}$. Then $y + C \cap \langle\langle X \rangle\rangle \neq \emptyset$ and there exist $z \in y + C \cap \langle\langle X \rangle\rangle$ and $z + C \in \pi_1(y + C) \cap \pi_1\langle\langle X \rangle\rangle = (y + C) \cap \langle \pi_1(X) \rangle >$. But $z \in y + C \iff z + C = y + C$. Thus $y + C \in \langle \pi_1(X) \rangle$ and $L/C = \langle \pi_1(X) \rangle$.

($\Leftarrow$) Assume $\langle \pi_1(X) \rangle \neq L/C$. Then $N = \overline{\langle\langle X \rangle\rangle} \quad L$ and $(\langle\langle X \rangle\rangle + C)/C \subseteq (N + C)/C \subseteq L/C$. But $(\langle\langle X \rangle\rangle + C)/C = \langle \pi_1(X) \rangle = L/C$ so $(N + C)/C = L/C$.

Now for large enough k, we have $N + C^k \quad L$. For if not then $N + C^k = L$ for every k and $N = \overline{N} = \cap_{k=0}^{\infty}(N + C^k) = L$, a contradiction.

Choose such a k. Then $(\langle\langle X \rangle\rangle + C^k)/C^k \subseteq (N + C^k)/C^k \quad L/C^k$. Since L/C^k is a finite dimensional vector space, $(N + C^k)/C^k \subseteq M/C^k$ where M is maximal in L containing C^k. But L/C^k is nilpotent, so by (**)

$$C/C^k = [L/C^k, L/C^k] \subseteq M/C^k.$$

However $\pi_1^i \pi_i = \pi_1$ implies $L/C = N + C/C \subseteq M/C \subseteq L/C$, a contradiction.

We now return to the main goal. That is, to find a faithful finite dimensional linear representation of L. Recall that

$$L = \mathbf{C}[[t]]^n \times \mathbf{C}[t]/(t^{d_1}) \times \cdots \times \mathbf{C}[t]/(t^{d_r}).$$

Denote the torsion submodule $\mathbf{C}[t]/(t^{d_1}) \times \cdots \times \mathbf{C}[t]/(t^{d_r})$ of L by T. It is easily shown that this is an ideal of L. In the theory of pro-p groups DDMS use Ado's theorem (Theorem 0.6) to get a faithful finite dimensional linear representation of a subgroup G_m of G into $\mathrm{Gl}_n(\mathbf{Q}_p)$. Because of the torsion T of L we are unable to do this. In lieu of this we consider $M = L/T$ with $\bar{t} : M \to M$ the induced linear operator and show that this is a torsion free powerful pro(finite dimensional)nilpotent Lie algebra.

In order to use Ado's Theorem we need the following taken from Commutativer Algebra by Atiyah and MacDonald [1, Prop. 3.5, p.39].

Proposition 5. *Let M be an A–module. Then the $S^{-1}A$ modules $S^{-1}M$ and $S^{-1}A \otimes M$ are isomorphic.*

It is clear that $M = L/T = $ bf $\mathrm{C}[[t]]^n$ is a $\mathbf{C}[[t]]$– module so by the previous theorem

$$M\left[\frac{1}{t}\right] = \mathbf{C}[[t]]\left[\frac{1}{t}\right] \otimes_{\mathbf{C}[[t]]} M = \mathbf{C}((t)) \otimes_{\mathbf{C}[[t]]} \mathbf{C}[[t]]^n$$

$$= (\mathbf{C}((t)) \otimes_{\mathbf{C}[[t]]} \mathbf{C}[[t]])^n = \mathbf{C}((t))^n$$

and $M[\frac{1}{t}]$ is a $\mathbf{C}((t))$–module.

Since M is a Lie algebra over $\mathbf{C}[[t]]$, it is clear that $M[\frac{1}{t}]$ is a Lie algebra over $\mathbf{C}[[t]][\frac{1}{t}] = \mathbf{C}((t))$.

We now give the main result.

Theorem 6. *There exists a faithful finite dimensional linear representation of $t^k M$ into $\mathrm{gl}_n(\mathbf{C}[[t]])$.*

Proof. By Ado's Theorem there exists a faithful finite dimensional linear representation of $M[\frac{1}{t}]$ into $\mathrm{gl}_n(\mathbf{C}((t)))$. Since M injects into $M[\frac{1}{t}]$ we have a faithful finite dimensional linear representation of M into $\mathrm{gl}_n(\mathbf{C}((t)))$.

Since M lies inside $\mathrm{gl}_n(\mathbf{C}((t)))$ as a finitely generated $\mathbf{C}[[t]]$–module of rank n, there exists $A_1, A_2, \ldots, A_n \in \mathrm{gl}_n(\mathbf{C}((t)))$ which generate M. Since $\mathbf{C}((t)) = \{\sum_{i=m}^{\infty} a_i t^i \ verta_i \in \mathbf{C}, m \in \mathbf{Z}\}$, choose k large enough so that $t^k A_j \in \mathbf{C}[[t]]$ for $1 \leq j \leq n$. Then $t^k M \hookrightarrow \mathrm{gl}_n(\mathbf{C}[[t]])$ is a faithful finite dimensional linear representation of $t^k M$ into $\mathrm{gl}_n(\mathbf{C}[[t]])$.

References

[1] Atiyah, M. F., Macdonald, I. G., *Introduction to commutative algebra*, Addison-Wesley Publishing Company, Inc., 1969.

[2] Bauerle, G. G. A., de Kerf, E. A., *Lie Algebras*, North–Holland Publishing Company, Amsterdam, 1990.

[3] Dixon, J. D., Du Sautoy, M. P. F., Mann, A., Segal, D. *Analytic pro-p Groups*, Cambridge University Press, 1991.

[4] Dugundji, J., *Topology*, Wm. C. Brown Publishers, Dubuque, Iowa, 1989.

[5] Higgins, P. J., *Introduction to Topological Groups*, Cambridge University Press, 1974.

[6] Humphreys, J. E. *Introduction to Lie Algebras and Representation Theory*, Springer-Verlag, New York, 1972, 1980.

[7] Hungerford, T. W., *Algebra*, Springer-Verlag, New York, 1987.

[8] Jacob, B., *Linear Algebra*, W. H. Freeman and Company, New York, 1990.

[9] Jacobson, N., *Lie Algebras*, Dover, New York, 1979.

[10] Lang, S., *Algebra*, Addison–Wesley Publishing Company, Inc., 1984.

[11] Magid, A. R., Identities for Prounipotent Groups, preprint.

[12] Matsumura, H., *Commutative ring theory*, Cambridge University Press, 1990.

[13] Munkres, J. R., *Topology*, Prentice–Hall, Inc., Englewood Cliffs, New Jersey, 1975.

Idealizers of Fractal Ideals in Free Group Algebras

AMNON ROSENMANN   University of Essen, Essen, Germany

Abstract

We examine the property of two-sidedness as well as the eigenrings of fractal ideals which are formed from the augmentation ideal in free group algebras. This is done by considering the structure of the trees which are associated to these right ideals.

1   Introduction

Let R be a ring with a unit element (denoted by 1) and let $\{y_i\}$ be a set of elements of R which freely generate a right ideal I (i.e. the $\{y_i\}$ form a basis for I as a right R-module). This is the setting needed for the construction of some special right ideals, called *fractal ideals*, which were introduced in Rosenmann and Rosset (1991) (see also Rosenmann, 1994). These are right ideals which are characterized through their construction by manner of a repeated duplication of the given original right ideal I, in a process that can be infinite. To understand the structure of a fractal ideal we associate to it a directed tree T. The nodes of T are denoted by products of the form $y_{i_1} \cdots y_{i_k}$. The root of T is denoted by 1, its direct descendants are the y_i, and each node $y = y_{i_1} \cdots y_{i_k}$ is either a leaf or it has outgoing edges to the nodes yy_i. There are now two ways to define inductively a (right) fractal ideal associated to such a tree T, and they relate to an increasing or to a decreasing fractal ideal. The increasing ideal is constructed by starting with the null ideal and at the n-th stage *adding* the right ideal generated by the (elements appearing as) *leaves* of T of level (depth) n (and we notice that the sum is direct). The decreasing ideal, on the other hand, is defined by starting with R, at the next stage we have I, and at the n-th stage we have the right ideal generated by the *nodes* of T of level less than or equal to n. In other words, in the decreasing case each direct summand yR, where y is an inner node of T of depth $n-1$, is replaced at the n-th stage by the right ideal $\sum_i yy_i R$, which is a duplication of $\sum_i y_i R$. If T is of finite depth then the associated fractal ideal is the one freely generated by the leaves of T, in both the increasing and the decreasing case. If T is of infinite depth then in the increasing case the fractal ideal is defined to be the *union* of the right ideals constructed at each stage. This also results in the right ideal freely generated by the leaves of T. In the decreasing case the fractal ideal is defined to be the *intersection* of the right ideals constructed at each stage. The decreasing fractal ideal always contains the increasing one, when both correspond to the same tree, but when the tree is of infinite depth the inclusion may be strict (see Rosenmann, 1994).

In this paper we examine questions concerning the idealizers of fractal ideals which are formed from the augmentation ideal in free group algebras. In section 2 we treat the property of two-sidedness, i.e. the case where the idealizer is the whole algebra. We show that the maximal two-sided ideal contained in a (right) decreasing fractal ideal associated with a tree T is also a decreasing fractal ideal, associated with a tree which is the "suffix-closure" of T. This enables us to answer e.g. the following (and also other) questions by considering the structure of the tree of a fractal ideal I_T:
  i) When is I_T a two-sided ideal?
  ii) When does I_T contain a non-trivial two-sided ideal?
For example, I_T can even be essential but without any non-trivial two-sided ideal within it.

In section 3 we give the structure of the eigenrings of special fractal ideals whose trees are minimal single-ended trees (i.e. with only one infinite path). Each such tree can be characterized by an infinite sequence d of integers, and the different types of the eigenrings are in 1-1 correspondence with the 3 different types of the sequences d: (i) constant, (ii) periodic but not constant, and (iii) non-periodic.

Finally, we would like to remark that because the fractal ideals we are investigating have free generators of the form $(x_i - 1)$ (or x_i in free algebras) it is easier to examine these ideals using their Schreier transversals (see Lewin, 1969) than right ideals in general, and to find nice properties they possess.

2 Two-sidedness of Fractal Ideals

Let R be a free group algebra KG, where K is a field and G is a free group generated by $X = \{x_1, \ldots, x_n\}$, $n \geq 2$. Let also I be the augmentation ideal of R with the generating set $\{y_1 = (x_1 - 1), \ldots, y_n = (x_n - 1)\}$. Then I is freely generated as a right ideal by $y_1, \ldots, y_n$ and we can construct fractal ideals with it. For our purpose here of investigating the property of two-sidedness of these (right) fractal ideals it is sufficient to examine only the *decreasing* ones. The reason is that when a tree contains a path which is *eventually-constant* (i.e. a path in which all edges, except possibly for finitely-many of them, go at the same direction y_i, for some fixed i), then the associated increasing fractal ideal is inessential (see Rosenmann, 1994) and therefore cannot be two-sided (since R is without zero divisors). On the other hand, when a tree does not contain eventually-constant paths, then it can be shown that the increasing and decreasing fractal ideals which correspond to that tree are identical. This can be done by using the following filtration on R (as in Rosenmann and Rosset, 1991, and in Rosenmann, 1994). Let

$$B_0 = \{(1 - x_{i_1}) \cdots (1 - x_{i_m}) \mid 1 \le i_j \le n, \, m \ge 0\}, \tag{1}$$

where the empty product is defined to be 1. Then for $k \ge 1$, let

$$B_k = \{\beta x_i^{-1} \gamma \mid \beta \in B_0, \, \gamma \in B_{k-1}, \, 1 \le i \le n\},$$

$$\tilde{B}_k = \bigcup_{j=0}^{k} B_j,$$

$$R_k = sp(\tilde{B}_k),$$

where the span is over K. Then

$$R = \bigcup_{k=0}^{\infty} R_k. \tag{2}$$

When the tree T is without eventually-constant paths and I_T is the associated decreasing right fractal ideal, then by repeated use of the simple identities

$$x_i^{-1} = 1 - (x_i - 1)x_i^{-1} = 1 - y_i x_i^{-1}, \tag{3}$$

we get that for each $\alpha \in R$

$$\alpha \equiv \beta \pmod{I_T}, \tag{4}$$

for some $\beta \in R_0$. This implies that I_T is (freely) generated by the elements represented by the labels at the leaves of T, the same as with the associated increasing right fractal ideal (and the elements represented by the inner nodes of T form a *Schreier transversal* for I_T in the algebra R). So in what follows, whenever we speak in this section of a fractal ideal, we mean a decreasing right fractal ideal, formed from the augmentation ideal I of R (and whenever we refer to a two-sided ideal the words two-sided will be present).

Next we present some notation and definitions. Every tree T can be defined by the set of labels $y_{i_1} \cdots y_{i_k}$ at its nodes. Such a set represents a tree if and only if it is *prefix-closed*, i.e. it contains every initial (left) segment of every element in it. The closure of T with respect to taking also suffixes (right segments) is denoted by $\overline{T}$. This is equivalent to saying that every segment (or sub-string) of every element of T is also in T. When T is both prefix and suffix-closed we say it is a *two-sided tree*. Thus $\overline{T}$ is two-sided.

As above, the fractal ideal associated with a tree T is denoted by I_T.

Theorem 2.1 *Let I_T be a fractal ideal of R. Then the maximal two-sided ideal contained in I_T is $I_{\overline{T}}$.*

Proof. First we will show that $I_{\overline{T}}$ is a two-sided ideal. For every $r = 1, 2, \ldots$, let $\overline{T}_r$ be the tree consisting of the nodes of $\overline{T}$ of depth $\leq r$. Then

$$I_{\overline{T}_r} = I_{\overline{T}} + I^r \tag{5}$$

and

$$I_{\overline{T}} = \bigcap_{r=1}^{\infty} I_{\overline{T}_r}. \tag{6}$$

It is then sufficient to show that $I_{\overline{T}_r}$ is two-sided for each r. $I_{\overline{T}_r}$ is a right ideal of finite codimension, freely generated by some $\beta_1, \ldots, \beta_m \in B_0$. We need then to show that for every $\alpha \in R$ and for every i, $1 \leq i \leq m$, $\alpha\beta_i \in I_{\overline{T}_r}$. The crucial point is that by using the identities (3), we may assume $\alpha \in B_0$. But in this case, if $\alpha\beta_i \notin I_{\overline{T}_r}$ then $\alpha\beta_i$ is an inner node of $\overline{T}_r$. Since $\overline{T}$, and thus also $\overline{T}_r$, is suffix-closed, β_i must be an inner node of $\overline{T}_r$. This, however, contradicts the fact that $\beta_i \in \overline{T}_r$. Therfore $I_{\overline{T}_r}$ is two-sided for each r and $\overline{T}$ is also a two-sided ideal.

Next we will show that $I_{\overline{T}}$ is the maximal two-sided ideal contained in I_T. Since $T \subseteq \overline{T}$ then clearly $I_{\overline{T}} \subseteq I_T$. Suppose now there is a two-sided ideal $J \subseteq I_T$ which properly contains $I_{\overline{T}}$. Then there is some $\alpha \in J \setminus I_{\overline{T}}$, and also there is some $r > 1$ such that $\alpha \notin I_{\overline{T}_r}$. Again, by using the identities (3), we get that there is $\eta \in R_0$ such that $\alpha \equiv \eta \pmod{I_{\overline{T}_r}}$. Let $\beta \in B_0$ be a term appearing in η. β is an inner node of $\overline{T}_r$, and therefore there exists $\gamma \in B_0$ such that $\gamma\beta$ is an inner node of T. Let $s = l(\gamma) + l(\beta) + 1$, where l denotes the length (with respect to the generators y_i). Let also $\zeta \in R_0$ be such that $\zeta \equiv \gamma\alpha \pmod{I_{T_s}}$, where T_s consists of the nodes of T of depth $\leq s$. Since $\gamma\beta$ is a term appearing in ζ and since it is an inner node of T_s, we get that $\gamma\alpha \notin I_{T_s}$. But $I_{T_s} \supseteq I_T \supseteq J$ and therefore $\gamma\alpha \notin J$. This contradicts the two-sidedness of J, and the proof is complete. $\qquad\square$

The following are immediate corollaries of Theorem 2.1.

Corollary 2.2 *I_T is a two-sided ideal if and only if $T = \overline{T}$.*

We say that T is the *full tree* of degree n if each of its nodes has n out-going edges.

Corollary 2.3 *I_T contains a non-trivial two-sided ideal if and only if $\overline{T}$ is not the full tree of degree n.*

Example 2.4 In figure 1 we see a tree T which represents a decreasing fractal ideal I_T, where G is a free group of rank 2 and, as before, I is the augmentation ideal of a group algebra $R = KG$ (a triangle in the figure with a number "i"

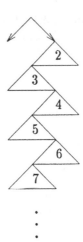

Figure 1: A tree of an essential fractal ideal which does not contain a two-sided ideal.

inside it represents the full tree of depth i, which corresponds to the i-th power of the augmentation ideal). It can be shown that I_T is an essential right ideal (see Rosenmann, 1994, for details about essentiality). On the other hand, I_T does not contain a two-sided ideal since $\overline{T}$ is the full tree of degree 2. The KG-module $A = KG/I_T$ gives then a faithful representation of the residually finite algebra KG, with the property that every element of A is annihilated by an essential right ideal of KG. This example also shows that the Martindale ring of quotients (see Passman, 1989, p. 83) is strictly included in the maximal ring of quotients of KG.

Corollary 2.5 *Let I_T be a fractal ideal and let $\underline{T}$ be the maximal two-sided tree contained in T. Then $I_{\underline{T}}$ is the minimal two-sided fractal ideal of I which contains I_T.*

Proof. This follows from the fact that for any trees T, T' one has $T \subseteq T'$ if and only if $I_T \subseteq I_{T}'$ and by Corollary 2.2. $\square$

We define T to be the *union* of the trees T_i when the set of labels on the nodes of T is the union of the sets of labels of the trees T_i.

Theorem 2.6 *Let J be a two-sided ideal contained in the augmentation ideal I of R. Then the minimal fractal ideal which contains J is a two-sided ideal.*

Proof. By Lemma 4.1 in Rosenmann (1994), the intersection of fractal ideals I_{T_i} is again a fractal ideal, such that if $T = \bigcup_i T_i$ then

$$I_T = \bigcap_i I_{T_i}. \tag{7}$$

Let I_T be the minimal fractal ideal which contains J, that is by the above

$$I_T = \bigcap_{I_{T_i} \supseteq J} I_{T_i}. \tag{8}$$

Let $L \supseteq J$ be the maximal two-sided ideal contained in I_T. By Theorem 2.1 L is a fractal ideal and by the minimality of I_T, $L = I_T$. $\square$

To conclude this section we give a result which applies to fractal ideals formed from the augmentation ideal $\sum_{i=1}^{n} x_i R$ of the free associative algebra $R = K\langle x_1, \ldots, x_n \rangle$. For this result, which concerns minimal sets of generators (i.e. such that every proper subset is not a generating set) of two-sided fractal ideals, we use the fact that, unlike the case of free group algebras, the generators x_i here do not have inverses.

Let $\leq$ be the following partial order on the set of words in the alphabet $X = \{x_1, \ldots, x_n\}$: $u \leq v$ if and only if u is a suffix of v. The set of leaves of T is denoted by $\mathcal{D}(T)$.

Theorem 2.7 *Let I_T be a two-sided fractal ideal of $R = K\langle x_1, \ldots, x_n \rangle$. Then the set of minimal elements of $\mathcal{D}(T)$, relative to the suffix partial order, forms a minimal set of generators of I_T.*

Proof. In $K\langle x_1, \ldots, x_n \rangle$ the increasing and decreasing fractal ideals which correspond to the same tree are identical. Then $\mathcal{D}(T)$ is a set of free generators of I_T as a right ideal. The rest of the proof follows from the fact that when u is a word in the alphabet X then for every non-zero elements $\alpha_1, \alpha_2 \in R$, each term in $\alpha_1 u \alpha_2$ contains u as a sub-word. $\square$

Another thing that can be proved is that the minimal two-sided ideal which contains a fractal ideal I_T of $K\langle x_1, \ldots, x_n \rangle$ is the fractal ideal which corresponds to the maximal two-sided tree contained in T.

3   Eigenrings of Fractal Ideals

The idealizer $\mathcal{I}(J)$ of a right ideal J of an algebra R is defined to be

$$\mathcal{I}(J) := \{r \in R \mid rJ \subseteq J\}. \tag{9}$$

$\mathcal{I}(J)$ is an algebra which contains J, and J is a two-sided ideal of it.

Let us first examine the idealizers of fractal ideals which are formed from the augmentation ideal $I = \sum_{i=1}^{n} x_i R$ of a free algebra $R = K{<}x_1, \ldots, x_n{>}$, $n \geq 2$. We use the elements $x_1, \ldots, x_n$ as free generators for I as a right ideal. The results of the previous section apply also to free algebras, but as for idealizers of fractal ideals, the picture is much more complicated in the case of free group algebras due to the existence of inverses. It is quite similar to what happens when examining the property of essentiality. In fact, the increasing and decreasing fractal ideals in $K{<}x_1, \ldots, x_n{>}$ which correspond to the same tree are identical (and the elements represented by the labels at the leaves of the tree form a set of free generators for the fractal ideal). But in the free group algebras there may be a great difference between the increasing and decreasing fractal ideals in connection with the essentiality property (see Rosenmann, 1994).

We use the following notation. X^* denotes the set of words in the alphabet X (or the set of monic monomials of R). $\overset{\circ}{T}$ denotes the set of inner nodes of T. When T is a tree and $u \in X^*$ we define the tree $T(u)$ to be

$$T(u) := \{v \in X^* \mid uv \in T\}. \tag{10}$$

Corollary 2.2 says that a fractal ideal I_T is a two-sided ideal if and only if $T(u) \subseteq T$ for every monomial u. The following proposition generalizes this result, noticing that a right ideal is two-sided if and only if its idealizer is the whole algebra.

Proposition 3.1 *Let I_T be a fractal ideal constructed from the augmentation ideal $I = \sum_{i=1}^{n} x_i R$ of a free algebra $R = K{<}x_1, \ldots, x_n{>}$. Then*

$$\mathcal{I}(I_T) \cong I_T + sp(\{u \in \overset{\circ}{T} \mid T(u) \subseteq T\}). \tag{11}$$

(sp is the linear span over K)

Proof. $T(u) \subseteq T$ if and only if $I_T \subseteq I_{T(u)}$, and this is equivalent to saying that $u\alpha \in I_T$ for every $\alpha \in I_T$, i.e. that $u \in \mathcal{I}(I_T)$. On the other hand, if $u \in X^*$ is not in $\mathcal{I}(I_T)$ and appears as a term in some $\beta \in R$, then $\beta \notin \mathcal{I}(I_T)$. This is because there exists $v \in I_T \cap X^*$ such that uv is an inner node of T, and this term appears also in βv. $\qquad \square$

The eigenring $\mathcal{E}(J)$ of a right ideal J is defined to be

$$\mathcal{E}(J) := \mathcal{I}(J)/J. \tag{12}$$

In special cases the eigenring of a right ideal of a finitely generated free algebra has a simple structure. For example, by a theorem of G. Bergman, the eigenring of a right ideal J of $K<X>$ which is generated by a single element is a finite dimensional commutative K-algebra (see Dicks, 1985). (In fact, by looking at a Schreier transversal tree for J we see that in this case it has only one "exit" towards the generator of J, and the leading terms in the elements of $\mathcal{I}(J)$ (mod J) must appear on the path from the root to this exit.) Analyzing the structure of the eigenring of a fractal ideal of $R = K<x_1,\ldots,x_n>$ may lead to interesting results. It is either K or has the structure of a quotient algebra of a free associative algebra $K<Z>$ (where Z may be infinite), with defining relations of the form

$$z_{i_1}\cdots z_{i_r} = 0 \tag{13}$$

or

$$z_{i_1}\cdots z_{i_r} = z_{j_1}\cdots z_{j_s}, \tag{14}$$

where the elements z_i, z_j are in Z. Assume now that R is again the free group algebra and I the augmentation ideal of it. We examine this special case: we look at fractal ideals whose trees are single-ended with the property that at each depth $i \geq 1$ there are exactly n nodes, $n-1$ of them are leaves and one is an inner node. Then to each such fractal ideal corresponds a sequence

$$d = (d_1, d_2, d_3, \ldots), \quad 1 \leq d_i \leq n, \tag{15}$$

which represents the end (infinite path) of the tree. We denote this fractal ideal by I_d.

Theorem 3.2 *Let I_d be a single-ended fractal ideal of $R = KG$. If d is constant then $\mathcal{E}(I_d) \cong K[z, z^{-1}]$, if d is periodic but not constant then $\mathcal{E}(I_d) \cong K[z]$, and if d is not periodic then $\mathcal{E}(I_d) \cong K$.*

Proof. Let $1 = u_0, u_1, u_2, \ldots$ be the terms (elements of B_0) appearing along the infinite path of T, i.e. $u_i = x_{d_1}\cdots x_{d_i}$. If d is not eventually-constant then these terms can serve as a Schreier transversal for I_d (this is shown by using the identities (3)). If, in addition, d is not periodic then $T(u_i) = T$ only when $i = 0$ and therefore

$$\mathcal{I}(I_d) = I_d + sp(\{1\}) \tag{16}$$

and

$$\mathcal{E}(I_d) \cong K. \tag{17}$$

Suppose now that d is periodic (but not constant). Let r be the length of the period. Then

$$T(u_i) = T(u_j) \iff i \equiv j \;(mod\; r). \tag{18}$$

In this case

$$\mathcal{I}(I_d) = I_d + sp(\{u_{kr} \mid k = 0, 1, 2, \ldots\}). \tag{19}$$

Since $u_{kr} = u_r^k$, $k = 0, 1, 2, \ldots$, then

$$\mathcal{E}(I_d) \cong K[z], \tag{20}$$

the ring of polynomials in one variable over K.

We are now in the case where d is eventually-constant. If d is a constant sequence, $d = (l, l, l, \ldots)$, for some $1 \leq l \leq n$, then I_d is a two-sided ideal generated (as a two-sided ideal) by the elements $(x_i - 1)$, $i \neq l$, and $\{x_l^j \mid j = 0, \pm 1, \pm 2, \ldots\}$ forms a Schreier transversal for I_d (see Rosenmann, 1994). Therefore

$$\mathcal{E}(I_d) \cong K[z, z^{-1}]. \tag{21}$$

It remains to consider the case where d is eventually-constant but not constant. Then it is of the form $(d_1, d_2, \ldots, d_r, l, l, l, \ldots)$, where $d_r \neq l$ and $r \geq 1$. The set

$$S = \{1, u_1, u_2, \ldots, u_{r-1}\} \cup \{u_r x_l^j \mid j = 0, \pm 1, \pm 2, \ldots\} \tag{22}$$

forms a Schreier transversal for I_d. Suppose now that $\alpha \cong \beta \pmod{I_d}$ and $\beta \in sp(S)$. If $\beta \in sp(\{1, u_1, \ldots, u_{r-1}\})$ then the situation is similar to the case where d is not eventually-constant and not periodic, and thus $\alpha I_d \not\subseteq I_d$ whenever $\beta \notin sp(\{1\})$. If, on the other hand, β contains terms of the form $u_r x_l^j$, j an integer, then evidently this will remain the case also when multiplying β by a power of x_l. Therefore, $\alpha x_l^k \notin I_d$ for every k such that $x_l^k \in I_d$ (and there are infinitely many such k-s). We conclude that when d is eventually-constant but not constant (and by the above, whenever d is not periodic) then

$$\mathcal{E}(I_d) \cong K. \tag{23}$$

$\square$

References

[1] Dicks, W. (1985). On the cohomology of one-relator associative algebras, *J. Algebra*, **97**: 79-100.

[2] Lewin, J. (1969). Free modules over free algebras and free group algebras: the Schreier technique, *Trans. AMS*, **145**: 455-465.

[3] Passman, D. S. (1989). *Infinite crossed products*, Academic Press, N.Y.

[4] Rosenmann, A. (1994, to appear). Essentiality of fractal Ideals, *Inter. J. Algebra and Computation*.

[5] Rosenmann, A., and Rosset, S. (1991). Essential and inessential fractal ideals in the group ring of a free group, *Bull. LMS*, **23**: 437-442.

Elements of Trace Zero in Central Simple Algebras

MYRIAM ROSSET   Bar-Ilan University, Ramat Gan, Israel
SHMUEL ROSSET   Tel-Aviv University, Ramat-Aviv, Tel-Aviv, Israel

To Dan Zelinsky in friendship and appreciation

1   Introduction

In 1936 Shoda proved (see [3]) that over a field k of characteristic zero a matrix of trace zero in $M_n(k)$ is a commutator $ab - ba$ (henceforth denoted $[a, b]$). Shoda's theorem was extended to fields of positive characteristic by Albert and Muckenhoupt 1957 (see [1]).

Recently Amitsur and Rowen raised the question whether the Shoda-Albert-Muckenhoupt theorem ('Shoda's theorem' from now on) holds in central simple algebras in general. Recall that if A is a central simple k-algebra then for some separable extension K of k, $A \otimes_k K$ is isomorphic, over K, to the full matrix algebra $M_n(K)$ for some n (n is called the degree of A). If $a \in A$ then all the invariants of the matrix associated by the isomorphism to $a \otimes 1$ are independent of the isomorphism and are, therefore, invariants of a. Thus a has a trace, a determinant (called 'norm'), a characteristic polynomial whose coefficients are in k etc. An element of A whose characteristic polynomial has n distinct roots will be called *generic*. Note that if a is generic the algebra it generates, $k[a]$, is separable, maximal commutative, equal to its own commutant and its dimension over k is n.

Amitsur and Rowen proved (see [2]) that if $c \in A$ is of trace zero and

$a \in A$ is *any* generic element then there are $u, v, w \in A$ such that

$$(1) \qquad\qquad c = [a, u] + [v, w].$$

They gave two proofs of their result, one using Capelli polynomials, the other using so called Brauer factor sets.

In this paper we prove, by methods of linear algebra, a similar but somewhat better result. To explain it we need a definition: a *generic pair* in A is a pair a, b of generic elements that satisfy the additional condition that

$$k[a] \cap k[b] = k.$$

This is equivalent to requiring that

$$\dim_k(k[a] + k[b]) = 2n - 1.$$

We prove that if a, b is a generic pair and c is an element of trace zero then there are $x, y \in A$ such that

$$(2) \qquad\qquad c = [a, x] + [b, y].$$

This result improves on the Amitsur-Rowen result in the sense that (2) is a set of *linear equations* (the unknowns being x and y) whereas (1) is a nonlinear set of equations (the unknowns being u, v and w). Where has the nonlinearity gone? It re-appears when one tries to find generic pairs. However, our proof of the existence of generic pairs in A shows that in $A \times A$ the set of generic pairs is Zariski open. This means that any set of two independently chosen 'random' elements of A is, with 'probability 1', a generic pair. Thus it seems that if one wants to write c as a sum of commutators the best (and probably the only) *practical* procedure would be to choose a random pair a, b and try to solve the linear equations (2). The point is that one knows, in advance, that almost surely they will be solvable. By contrast if one chooses a single random element a and tries to solve the equation

$$c = [a, x]$$

then it is easy to show that, almost surely, it is *not* solvable. This may be why it is not known, yet, whether the extension of Shoda's theorem to central simple algebras is true or not.

2   Generic pairs

In this section we prove the existence of generic pairs in central simple algebras.

In order not to have to deal with some irrelevant special cases we will assume in this section, and indeed for the rest of this note, that the base field, k, is infinite.

Let A be a central simple k-algebra of degree n, i.e. of dimension n^2 over k and let $e_1, ..., e_m$ be a basis of A, so that $m = n^2$.

If

$$a = a_1 e_1 + \cdots + a_m e_m$$

and

$$b = b_1 e_1 + \cdots + b_m e_m$$

are two "general" elements of A we want to express the conditions that make the pair a, b generic. It is not hard to see that the coefficients of the characteristic polynomial of a are polynomials, with k coefficients, of its coordinates $a_1, ..., a_m$. The condition that a be generic is that the discriminant of its characteristic polynomial be non-zero. This discriminant is, in turn, a polynomial function (with integer coefficients) of the coefficients of the characteristic polynomial. So, altogether, the discriminant of the characteristic polynomial of a is a polynomial function of its coordinates. The condition that a be generic is that this discriminant does not vanish. Thus a is generic, except when it lies in the algebraic subset defined by the vanishing of the discriminant.

Suppose now that a and b are two 'independent' generic elements. Then the pair a, b is generic iff the $2n - 1$ elements

$$1, a, a^2, ..., a^{n-1}, b, b^2, ..., b^{n-1}$$

are linearly independent over k. Writing each of the elements in this list in terms of the basis $e_1, ..., e_m$ we get a $(2n - 1) \times m$ matrix, $\mathcal{M}$, whose entries are polynomial functions of the coordinates of a and b. It is clear that these $2n - 1$ elements are linearly independent iff the matrix, $\mathcal{M}$, expressing their dependence on the basis is of rank $2n - 1$, and this happens iff some $(2n - 1) \times (2n - 1)$ minor of $\mathcal{M}$ does not vanish.

Each of these minors is a polynomial function of the coordinates

$$a_1, ..., a_m, b_1, ..., b_m$$

of the elements a and b. The pair a, b is generic iff the discriminant functions, described above, for both a and b and one of these minors do not vanish (simultaneously). Since we are assuming that k is an infinite field it is enough to prove that, as polynomials in the 'variables' $a_1, ..., b_m$, the discriminants and at least one of these minors do not vanish identically. To do that it suffices to show that in some extension, K, of k there is, for each of them, a 'substitution' in K that gives a non-zero value. But this is equivalent to showing that in $A \otimes_k K$ there are generic pairs.

To show that for some K there are generic pairs in $A \otimes_k K$ we can follow several routes. One of them is to note that in a cyclic algebra there are ready to use generic pairs. To explain that recall that a cyclic algebra (over K) is defined by two elements a, b such that a generates a cyclic extension of K of degree n, b is invertible and conjugation by it preserves $K(a)$ and induces on it an automorphism that is a generator of the Galois group of $K(a)$ over K. By definition the powers $1, b, ..., b^{n-1}$ are independent over $K(a)$ and hence the subalgebra $K(b)$ is n-dimensional over K. Also b^n is central, commuting as it does with both a and b. Denoting b^n by β we see that the characteristic polynomial of b is $x^n - \beta$, which is separable if n is not divisible by the characteristic of K. We claim that, in this case, a, b is a generic pair. Indeed, it is clear that a and b are generic and the condition $K(a) \cap K(b) = K$ follows from the fact that the powers of b are linearly independent over $K(a)$.

If $\mathrm{char}(k)$ divides n the situation can be saved by the following consideration. It is easy to see that the powers, from 0 to $n - 1$, of $\lambda a + b$, where λ is an arbitrary (but fixed) element of K, are linearly independent over $K(a)$. So it is clear that, for every λ, $K(a) \cap K(\lambda a + b) = K$. If λ can be found such that $\lambda a + b$ is generic we will be done. It can be seen that the discriminant of the characteristic polynomial of $\lambda a + b$ is a non-trivial polynomial of λ so that, indeed, there are many values of λ that render $\lambda a + b$ generic, as desired.

It is well known that if A is a central simple k-algebra there are extensions, K, of k that 'make' A cyclic, i.e. such that $A \otimes_k K$ is cyclic over K. Thus, by the considerations above, there are (many) generic pairs in A.

3   The main result

In this section we prove

(3.1) Theorem. *If A is a central simple k-algebra and a, b is a generic pair in A then for every $c \in A$ which is of trace zero there exist $x, y \in A$ such that*

$$c = [a, x] + [b, y].$$

Let T be the bilinear form on A defined by

$$T(u, v) = \mathrm{tr}(uv).$$

It is a symmetric non-degenerate form on A. Indeed, to see that it is non-degenerate it suffices to show that its extension to $A \otimes_k K$ is non-degenerate for some K. But if $A \otimes_k K$ is a matrix algebra (i.e. if K splits A) then it is an easy exercise to show that the bilinear form $tr(uv)$ is non-degenerate.

If $f : A \to A$ is a k-linear endomorphism its adjoint, relative to T, is the linear transformation $f^* : A \to A$ satisfying, identically,

$$T(f(u), v) = T(u, f^*(v)).$$

If $f^* = f$ we say that f is symmetric, and if $f^* = -f$ we say that it is anti-symmetric. We claim that the map f_a defined by

$$f_a(x) = ax - xa,$$

where a is a fixed (but arbitrary) element of A, is anti-symmetric. This follows by a simple computation:

$$\mathrm{tr}((ax - xa)y) = \mathrm{tr}(xya - xay) = -\,\mathrm{tr}(x(ay - ya)).$$

We will use the so called "Fredholm alternative" of linear algebra which asserts that for every linear transformation $f : A \to A$

$$\mathrm{Im}(f) = (\ker(f^*))^{\perp}$$

where $(\cdot)^{\perp}$ denotes the perpendicular to $(\cdot)$. In particular for every $a \in A$, $\ker(f_a^*)$ which, by virtue of the anti-symmetry of f_a, is equal to $\ker(f_a)$, is equal to the commutant of a. When a is generic this commutant is $k[a]$, and is n-dimensional over k, so

$$\mathrm{Im}(f_a) = k[a]^{\perp}$$

and its dimension is $n^2 - n$ in this case.

To prove our theorem we will prove that the linear transformation

$$\varphi : A \oplus A \to A$$

defined by

$$\varphi(x, y) = [a, x] + [b, y]$$

has a 1-codimensional image in A. Denoting by A' the subspace of elements of trace zero we know that $\mathrm{Im}(\varphi) \subseteq A'$, so if it is of codimension one it must be equal to A'.

We can express φ as a composite $\psi \circ \theta$ where $\theta : A \oplus A \to A \oplus A$ is defined by

$$\theta(x, y) = ([a, x], [b, y])$$

and

$$\psi(u, v) = u + v.$$

Hence the dimension of $\mathrm{Im}(\varphi)$ is $\dim(\mathrm{Im}\theta)$-$\dim(\ker\psi')$ where ψ' is the restriction of ψ to $\mathrm{Im}\theta$. The dimension of $\mathrm{Im}\theta$ is, evidently, $2n^2$-$\dim(\ker\theta)$.

Now the kernel of θ is

$$\{(x, y) \in A \oplus A : [a, x] = 0, [b, y] = 0\} = k[a] \oplus k[b].$$

Thus the dimension of $\ker\theta$ is $2n$.

To compute the dimension of $\ker(\psi')$ note that $\mathrm{Im}\theta$ is the direct sum $[a, A] \oplus [b, A]$, so that the elements of $\ker(\psi')$ have the form $(u, -u)$ where

$$u \in [a, A] \cap [b, A].$$

In other words $\ker(\psi')$ is the anti-diagonal of $([a, A] \cap [b, A]) \oplus ([a, A] \cap [b, A])$. It follows that its dimension is equal to the dimension of $[a, A] \cap [b, A]$. But $[a, A]$ is just $\mathrm{Im}(f_a)$ where f_a is as defined above. So to compute the dimension of $\ker(\psi')$ we need to compute the dimension of $\mathrm{Im}(f_a) \cap \mathrm{Im}(f_b)$.

We observed that if a is generic $\mathrm{Im}(f_a) = k[a]^\perp$. From the formula

$$U^\perp \cap V^\perp = (U + V)^\perp$$

we deduce that

$$\mathrm{Im}(f_a) \cap \mathrm{Im}(f_b) = (k[a] + k[b])^\perp.$$

As the pair a, b is generic $\dim(k[a]+k[b])=2n-1$, and therefore the dimension of its perpendicular subspace is

$$n^2 - 2n + 1$$

which is thus also the dimension of $\ker(\psi')$.

Altogether we see that the dimension of $\operatorname{Im}(\varphi)$ is

$$2n^2 - 2n - (n^2 - 2n + 1) = n^2 - 1.$$

This concludes the proof.

Acknowledgement. We are much indebted to Louis Rowen for useful information and conversations.

References

[1] A.A. Albert and B. Muckenhoupt, On matrices of trace zero, Michigan Math. J. 3 (1957), 1-3.

[2] S.A. Amitsur and L.H. Rowen, Elements of reduced trace 0, Israel J. Mathematics (to appear)

[3] K. Shoda, Einige satze uber Matrizen, Japanese J. Math. 13 (1936), 361-365.

Canonical Modules and Factorality of Symmetric Algebras

Aron Simis[*]

Instituto de Matemática, Universidade Federal da Bahia

40170–210 Salvador, Bahia, Brazil

aron@brufba.bitnet

Bernd Ulrich[†]

Department of Mathematics, Michigan State University

East Lansing, Michigan 48824

21144bfu@msu.bitnet

Wolmer V. Vasconcelos[‡]

Department of Mathematics, Rutgers University

New Brunswick, New Jersey 08903

vasconce@rings.rutgers.edu

1 Introduction

Symmetric algebras of modules are basically families of polynomial rings. More precisely, let R be a ring and let E be a module with a presentation

$$R^m \xrightarrow{\varphi} R^n \longrightarrow E \longrightarrow 0.$$

The corresponding algebra presentation of the symmetric algebra of E, $S = S(E)$, is

$$S = R[T_1, \ldots, T_n]/(\mathbf{f}),$$

---

[*]Partially supported by CNPq, Brazil.

[†]Partially supported by the NSF.

[‡]Partially supported by the NSF.

where $\mathbf{f}$ is a set of linear forms in the T_i's with coefficients in R:

$$\mathbf{f} = [f_1, \ldots, f_m] = [T_1, \ldots, T_n] \cdot \varphi.$$

Despite the simplicity of this description, and unlike polynomial rings (when E is a free R–module), it has been rather untractable to determine even simple issues such as when $S(E)$ is an integral domain.

Here we consider a more specialized task, deciding when $S(E)$ is a factorial domain. This requires that R be factorial and that all the symmetric powers of E, $S_r(E)$, be reflexive modules (cf. [11]). It is this infinity of ancillary conditions that lies at the root of the difficulty. If R is a Cohen–Macaulay ring, one condition that expresses partially these requirements is found in the following family of bounds on the sizes of the Fitting ideals of the module E. Let $I_t(\varphi)$ be the ideal generated by the minors of order t of a presentation matrix φ of E. The conditions

$$\mathcal{F}_k : \qquad \text{height } I_t(\varphi) \geq \text{rank}(\varphi) - t + 1 + k, \ 1 \leq t \leq \text{rank}(\varphi),$$

play various roles in the ideal theory of the algebra $S(E)$. If $S(E)$ is a domain, $\mathcal{F}_1$ is satisfied, while if $S(E)$ is factorial, the condition $\mathcal{F}_2$ holds ([6, p. 665]). On the other hand, when R is factorial and E is a module of projective dimension one, $\mathcal{F}_2$ is precisely the required global condition to make $S(E)$ factorial ([1]). Note that in this case $S(E)$ is a complete intersection. On the other hand, condition $\mathcal{F}_2$ cannot occur in modules of projective dimension two, but may occur in modules of higher projective dimensions ([6]), although none of these examples leads to factorial algebras.

This state of affairs is the motivation for our study here that settles some cases of interest of when $S(E)$ is factorial. Actually, our approach turns on a more general one, the description of the canonical module of $S(E)$. This provides some obstructions for the symmetric algebra $S(E)$ to be Gorenstein. The central question concerns the following:

Conjecture 1.1 (Strong Factorial Conjecture) If $S(E)$ is normal and quasi–Gorenstein, then it is a complete intersection.

This conjecture is stated much too broadly. There is more ground for one of its instances ([6]):

Conjecture 1.2 (Factorial Conjecture) Let R be a regular local ring. If $S(E)$ is factorial, then it is a complete intersection.

The setting for most of our discussion will be of modules over a regular local ring $(R, \mathfrak{m})$. Technically, the issue is to estimate, for a module E, the depth of the local ring obtained by localizing $S = S(E)$ at the prime ideal $\mathfrak{m}S$. One of our main results (Theorem 2.1), describes conditions that permit the resolution of the conjectures in some relatively general cases (Theorem 2.2 and its corollaries), and some others in dimension at most five (Theorem 3.1). Nevertheless, in full generality these questions remain unresolved.

2   Canonical Module

Our main technical tool to study these conjectures is the following:

Theorem 2.1 *Let* $(R, \mathfrak{m})$ *be a regular local ring, let* E *be a finitely generated* R*–module, and write* $S = S(E)$. *Assume*

  (i) E *satisfies* $\mathcal{F}_0$;

  (ii) *for every* $\mathfrak{q} \in \mathrm{Ass}(S_{\mathfrak{m}S})$, *proj dim* $E_{\mathfrak{q} \cap R} \leq 1$;

  (iii) *for every* $\mathfrak{q} \in \mathrm{Spec}(S_{\mathfrak{m}S})$ *with depth* $S_{\mathfrak{q}} \leq 1$, $S_{\mathfrak{q}}$ *is a complete intersection;*

  (iv) S *is quasi–Gorenstein.*

If proj dim $E \geq 2$, *then* depth $S_{\mathfrak{m}S} \geq 2$.

Proof. Consider an exact sequence

$$0 \to L \longrightarrow R^n \longrightarrow E \to 0 \tag{1}$$

with rank $L = \ell$. Notice that $\mathrm{Hom}_R(\wedge^\ell L, R) \simeq R$. For the natural map

$$\epsilon : \mathrm{Hom}_R(\wedge^\ell L, R) \otimes_R S \longrightarrow \mathrm{Hom}_R(\wedge^\ell L, S),$$

we are going to prove that $\epsilon \otimes_S S_{\mathfrak{m}S}$ is an isomorphism.

To do so, notice that (1) gives rise to a homogeneous presentation $S \simeq B/J$, where $B = S(R^n) = R[T_1, \ldots, T_n]$. Because of (i), grade $J = \ell$ ([10, Theorem 2.6]), and hence (iv) yields a homogeneous isomorphism

$$S(\ell) \quad \sim \quad \mathrm{Ext}_B^\ell(S, B). \tag{2}$$

Quite generally, there is a natural graded homomorphism by [3, (3.1)],

$$\varphi : \mathrm{Ext}_B^\ell(S, B) \longrightarrow \mathrm{Hom}_B(\wedge^\ell J, S), \tag{3}$$

which becomes an isomorphism upon localizing at any multiplicative set where J is generated by a regular sequence of length ℓ ([3, Proposition 4]). Thus by (ii) and (iii), $\varphi \otimes_S S_{\mathfrak{q}}$ is an isomorphism for every $\mathfrak{q} \in \mathrm{Spec}(S_{\mathfrak{m}S})$ with depth $S_{\mathfrak{q}} \leq 1$ and therefore $\varphi \otimes_S S_{\mathfrak{m}S}$ is an isomorphism. Furthermore $\varphi \otimes_R R_{\mathfrak{p}}$ is a homogeneous isomorphism for every $\mathfrak{p} \in \mathrm{Spec}(R)$ with depth $R_{\mathfrak{p}} \leq 1$, and hence $[\varphi]_{-\ell}$ is an isomorphism as well.

The exact sequence (1) yields a graded epimorphism

$$L \otimes_R S(-1) \longrightarrow J \otimes_B S,$$

which, by (ii), is an isomorphism locally at every associated prime of $S_{\mathfrak{m}S}$, and which is an isomorphism in degree one. This map induces a homogeneous map

$$\psi : \operatorname{Hom}_B(\wedge^{\ell} J, S) \longrightarrow \operatorname{Hom}_R(\wedge^{\ell} L, S)(\ell), \tag{4}$$

with $\psi \otimes_S S_{\mathfrak{m}S}$ and $[\psi]_{-\ell}$ being isomorphisms. Now putting (2), (3), and (4) together, we obtain a homomorphism of graded S–modules

$$\chi : S \longrightarrow \operatorname{Hom}_R(\wedge^{\ell} L, S),$$

where $\chi \otimes_S S_{\mathfrak{m}S}$ and $[\chi]_0$ are isomorphisms. The latter condition gives a factorization

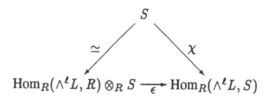

which shows that $\epsilon \otimes_S S_{\mathfrak{m}S}$ is an isomorphism.

Now consider a presentation

$$F_1 \longrightarrow F_0 \longrightarrow \wedge^{\ell} L \to 0, \tag{5}$$

where F_0 and F_1 are free R–modules of finite ranks, and dualize (5) to obtain the exact sequence

$$0 \to \operatorname{Hom}_R(\wedge^{\ell} L, R) \xrightarrow{\delta} \operatorname{Hom}_R(F_0, R) \longrightarrow \operatorname{Hom}_R(F_1, R). \tag{6}$$

Applying $\operatorname{Hom}_R(\bullet, S)$ to (5), and $\bullet \otimes_R S$ to (6), we obtain a commutative diagram

$$
\begin{array}{ccccccc}
0 & \longrightarrow & \operatorname{Hom}_R(\wedge^{\ell} L, S) & \longrightarrow & \operatorname{Hom}_R(F_0, S) & \longrightarrow & \operatorname{Hom}_R(F_1, S) \\
 & & {\scriptstyle\epsilon}\big\uparrow & & \big\| & & \big\| \\
0 & \longrightarrow & \operatorname{Hom}_R(\wedge^{\ell} L, R) \otimes_R S & \xrightarrow{\delta \otimes_R S} & \operatorname{Hom}_R(F_0, R) \otimes_R S & \longrightarrow & \operatorname{Hom}_R(F_1, R) \otimes_R S
\end{array}
$$

where the top row is exact. If we denote by $\mathbb{F}$ the bottom row, we have seen that $\epsilon \otimes_S S_{\mathfrak{m}S}$ is an isomorphism, and hence $\mathbb{F} \otimes_S S_{\mathfrak{m}S}$ is exact as well.

Now suppose that depth $S_{\mathfrak{m}S} \leq 1$. Then since $\mathbb{F} \otimes_S S_{\mathfrak{m}S}$ is an exact sequence of free $S_{\mathfrak{m}S}$–modules, $\delta \otimes_S S_{\mathfrak{m}S}$ would have to be split injective, and hence δ would split. Thus, dualizing (6) and comparing it to (5), one sees that $\wedge^{\ell} L$ would have a free summand. But then L itself would have a free summand of rank ℓ, hence L would be free, and proj dim $E \leq 1$. $\qquad\square$

Notice that assumptions (i) and (ii) of Theorem 2.1 are automatically satisfied if $S = S(E)$ happens to be a domain, and (i), (ii), and (iii) hold if S is normal.

Theorem 2.2 *Let $(R, \mathfrak{m})$ be a regular local ring, let E be a finitely generated R–module satisfying $\mathcal{F}_1$, and write $S = S(E)$. If S is a Gorenstein ring and $S_{\mathfrak{m}S}$ is a complete intersection, then S is a complete intersection.*

Proof. Set $k = \dim S_{\mathfrak{m}S}$. Since E satisfies $\mathcal{F}_1$ and S is Cohen–Macaulay, grade $\mathfrak{m}S = k \geq 1$ and S is a domain ([14, Proposition 3.3]). Let $\mathfrak{m} = (x_1, \ldots, x_d)$, let Z be a d by $k - 1$ matrix of indeterminates, and in $R' = R(Z)$, consider the elements

$$[y_1, \ldots, y_{k-1}] = [x_1, \ldots, x_d] \cdot Z.$$

Since grade $\mathfrak{m}S = k$, $y_1, \ldots, y_{k-1}$ form a regular sequence and generate a prime ideal in $S \otimes_R R'$ ([8]). Write $\overline{R} = R'/(y_1, \ldots, y_{k-1})R'$ and let $\overline{\mathfrak{m}}$ be the maximal ideal of $\overline{R}$. Then $\overline{S} = S \otimes_R \overline{R}$ is a Gorenstein domain and $\overline{S}_{\overline{\mathfrak{m}S}}$ is a one-dimensional complete intersection. Since on the other hand, $\overline{S} = S_{\overline{R}}(E \otimes_R \overline{R})$, Theorem 2.1 now implies that proj dim $_{\overline{R}}(E \otimes_R \overline{R}) \leq 1$. Thus $\overline{S}$ is a complete intersection, which gives the assertion. $\qquad\square$

Corollary 2.3 *Let R be a polynomial ring over a field, let E be a finitely generated graded R–module satisfying $\mathcal{F}_1$, and assume that all but at most four of the generating relations of E are linear. If $S(E)$ is Gorenstein, then it is a complete intersection.*

Proof. Let $\mathfrak{m}$ be the irrelevant maximal ideal of R. By Theorem 2.2 it suffices to show that $S(E)_{\mathfrak{m}S(E)}$ is a complete intersection if $S(E)$ is Gorenstein. However, our assumption implies that $E \simeq G/(a_1, \ldots, a_4)$, where G is an R–module with linear presentation matrix. Thus $S(G)_{\mathfrak{m}S(G)}$ is a regular local ring, and $S(G)_{\mathfrak{m}S(G)}/K \simeq S(E)_{\mathfrak{m}S}$ is Gorenstein with $\nu(K) \leq 4$. Now the assertion follows since every four generated Gorenstein ideal in a regular local ring has to be a complete intersection. $\qquad\square$

Our next two corollaries show that Conjectures 1.1 and 1.2 are equivalent in many cases.

Corollary 2.4 *Let R be a regular local ring, and let E be a vector bundle with proj dim $E \neq 1$. If $S(E)$ is normal and quasi–Gorenstein, then $S(E)$ is factorial.*

Proof. Notice that a normal symmetric algebra $S(E)$ is factorial if and only if E satisfies $\mathcal{F}_2$ ([6, Theorem 2.1]). Furthermore, since E is a vector bundle, the latter condition is equivalent to the inequality $\dim S(E)_{\mathfrak{m}S} \geq 2$, which holds by Theorem 2.1. $\qquad\square$

The next result is a generalization of [13, Corollary 2.9].

Corollary 2.5 *Let R be a regular local ring, let E be a finitely generated R–module, and write $S = S(E)$. If S satisfies (S_2) and is quasi–Gorenstein, and $S_{\mathfrak{p}}$ is factorial for every $\mathfrak{p} \in \mathrm{Spec}(S)$ with $\dim S_{\mathfrak{p}} \leq 3$, then S is factorial.*

Proof. We first notice that S is a normal domain. We may assume that E satisfies $\mathcal{F}_2$ locally on the punctured spectrum. Thus, if S is not factorial, then dim $S_{\mathfrak{m}S} = 1$, and hence by Theorem 2.1, S would be a complete intersection. But now [4, Corollaire 3.14] would imply that S is factorial because it is factorial locally in codimension three (alternatively, one can arrive at a contradiction using a presentation of S and the fact that $Cl(S) = \mathbb{Z}[\mathfrak{m}S]$ by [7]). $\qquad\square$

Corollary 2.6 *Let R be a regular local ring and let I be a normal ideal of linear type with height $I \geq 2$ that is normally torsion-free on the punctured spectrum of R. The following are equivalent:*

(a) $R[It]$ *is quasi–Gorenstein.*

(b) *height $I = 2$, I is height unmixed, and I is perfect or normally torsion–free.*

Proof. In the light of Theorem 2.1, it suffices to show that $R[It]$ is quasi–Gorenstein, if I is height unmixed of height two and I is normally torsion–free (which was done in [12, 3.4] for reduced ideals). Since I is normal, normally torsion–free of linear type, [7, 3.2.1] gives the following equality in the divisor class group of $R[It]$:

$$[\omega_{R[It]}] = \sum_{\mathfrak{p}\in P} n_{\mathfrak{p}}[T(\mathfrak{p})],$$

where P is the set of minimal primes of I and $T(\mathfrak{p})$ are the torsion primes defined in [10] and [7]. However, for every $\mathfrak{p} \in P$, $I_{\mathfrak{p}}$ is a complete intersection of height two, hence

$$0 = [\omega_{R[I_{\mathfrak{p}}t]}] = n_{\mathfrak{p}}[T(\mathfrak{p}_{\mathfrak{p}})].$$

Therefore $n_{\mathfrak{p}} = 0$, because $[T(\mathfrak{p}_{\mathfrak{p}})]$ has infinite order in the divisor class group of $R[I_{\mathfrak{p}}t]$ (e.g. [7, 3.2.1]). $\qquad\square$

Corollary 2.7 *Let R be a regular local ring and let E be a finitely generated R–module that is free locally in codimension 2. If $S(E)$ is normal and quasi-Gorenstein, then E is reflexive.*

Proof. We need to show that E satisfies Serre's condition (S_2). Let $\mathfrak{m}$ be a prime ideal of R with height $\mathfrak{m} \geq 3$. Localizing at $\mathfrak{m}$ we may assume that R is local with maximal ideal $\mathfrak{m}$. We will prove that depth $E \geq 2$; it suffices to consider the case proj dim $E \geq 2$. But then by Theorem 2.1, since $S = S(E)$ satisfies (S_2), grade $\mathfrak{m}S \geq 2$, which gives depth $E \geq 2$. $\qquad\square$

Corollary 2.8 *Let R be a regular local ring and let E be a finitely generated R module with proj dim $E = 2$. Then $S(E)$ cannot be normal and quasi–Gorenstein.*

First Proof. Suppose that $S = S(E)$ is normal and quasi–Gorenstein. Then we may assume, by induction on the dimension of R, that E has projective dimension at most one locally on the punctured spectrum of R. Furthermore, by Theorem 2.1, height $\mathfrak{m}S \geq 2$, and hence grade $\mathfrak{m}S \geq 2$ because S satisfies (S_2). Thus, by tensoring a resolution of E over R,

$$0 \to R^p \longrightarrow R^m \xrightarrow{\varphi} R^n \longrightarrow E \to 0$$

with S, we get the complex

$$0 \to R^p \otimes S \longrightarrow R^m \otimes S \longrightarrow R^n \otimes S \longrightarrow E \otimes S \to 0,$$

which is acyclic by the hypothesis and the acyclicity lemma. In particular we have

$$\mathrm{Tor}_1^R(E, S(E)) = 0,$$

and thus by [16, Proposition 3.2], E must have projective dimension one, which is a contradiction. $\square$

Second Proof. Let L denote the image of φ. There is a homomorphism from $L \otimes B$ onto the defining ideal J of $S(E)$ $(B = R[T_1, \ldots, T_n])$, and therefore a surjection

$$\beta : L \otimes S \longrightarrow J/J^2$$

of $S(E)$–modules of the same rank. But $L \otimes S$ is a torsion–free module of projective dimension 1 (as grade $\mathfrak{m}S \geq 2$), so β must be an isomorphism. By Gulliksen's Theorem ([2, Theorem 1.4.9]; see also [15]), $S(E)$ is then a complete intersection. $\square$

Remark 2.9 By a quirk, the first proof can be extended to modules whose third betti number is 1.

Proof. We keep the setting of the proof above. Let L be the module of first–order syzygies of E:

$$0 \to R \longrightarrow R^p \longrightarrow R^m \longrightarrow L \to 0.$$

Tensoring with $S(E)$, by the acyclicity lemma,

$$\mathrm{Tor}_i^R(L, S(E)) = 0, \text{ for } i \geq 1.$$

As in the proof of [16, Proposition 3.2], we obtain a contradiction by repeatedly pushing upwards the dimension of R. First note that, from $\mathrm{Tor}_i^R(L, E) = 0$ for $i \geq 1$, one obtains dim $R \geq 5$. On the other hand, since $\beta_3(E) = 1$, the Weyman's complex ([17]) over $S_2(E)$ has length 5 and is therefore acyclic as by assumption E has projective dimension at most 1 on the punctured spectrum. From $\mathrm{Tor}_i^R(L, S_2(E)) = 0$ for $i \geq 1$, we now get that dim $R \geq 7$. Because the length of the Weyman's complex over $S_t(E)$ is $2t + 1$, the argument ensures that dim R grows without bound. $\square$

Remark 2.10 The argument in Remark 2.9 can also be used to prove that the conjecture holds for modules whose third betti number is at most 2.

3   Dimension Five

We are going to consider the conjecture for dim $R = 5$.

Theorem 3.1 *Let R be a regular local ring of dimension at most five, let E be a finitely generated R-module, and write $S = S(E)$. If S is normal and quasi–Gorenstein, and $S_{\mathfrak{p}}$ is Cohen–Macaulay for every $\mathfrak{p} \in \operatorname{Spec}(S)$ with dim $S_{\mathfrak{p}} \leq 3$, then S is a complete intersection.*

Proof. Suppose the assertion is false, then, localizing at a suitable prime, we may assume that R is local with maximal ideal $\mathfrak{m}$, proj dim $E \geq 2$, and E has projective dimension at most one locally on the punctured spectrum of R. Now by Theorem 2.1, grade $\mathfrak{m}S \geq 2$. Thus depth $S_i(E) \geq 2$ for every i, and hence dim $R \geq 4$.

Consider a minimal presentation

$$0 \to L \longrightarrow R^n \longrightarrow E \to 0,$$

with depth $L \geq 3$. Furthermore, since height $\mathfrak{m}S \geq 2$, the dimension formula for symmetric algebras ([10, Theorem 2.6]) gives that rank $L \leq \dim R - 2 \leq 3$. Hence after adjoining free summands if needed, we may assume that E admits a free presentation with rank $L = 3$.

Now consider the augmented approximation complex of E,

$$\mathcal{Z}_{\bullet}: \qquad 0 \to L_3 \otimes B(-3) \longrightarrow L_2 \otimes B(-2) \longrightarrow L_1 \otimes B(-1) \longrightarrow B \longrightarrow S \to 0,$$

where $B = R[T_1, \ldots, T_n]$, $L_1 = L$, $L_2 = (\wedge^2 L)^{**}$, and $L_3 = (\wedge^3 L)^{**} \simeq R$ ([6]). Since L is free on the punctured spectrum and dim $R \geq 3$, $\mathcal{Z}_{\bullet}$ is exact. In particular depth $L_2 \geq 3$, because depth $S_2(E) \geq 2$ and depth $L_1 \geq 3$. But then, again using the approximation complex and the fact that S is Cohen–Macaulay in codimension 3, one concludes that

$$2 \operatorname{depth} S_{\mathfrak{p}} \geq \dim S_{\mathfrak{p}} + 2$$

for every $\mathfrak{p} \in \operatorname{Spec}(S)$ with dim $S_{\mathfrak{p}} \geq 2$. This together with the quasi–Gorenstein property of S implies that S is Cohen–Macaulay (cf. [9, Lemma 5.8] or [5, Theorem 1.6]), and hence Gorenstein. Since furthermore the approximation complex is exact, we may now use [6, Corollary 6.8] to conclude that proj dim $E \leq 1$. $\square$

References

[1] L. Avramov, Complete intersections and symmetric algebras, J. Algebra **73** (1980), 249–280.

[2] T. H. Gulliksen and G. Levin, *Homology of Local Rings*, Queen's Papers in Pure and Applied Math., No. **20**, Queen's University, Kingston, 1969.

[3] A. Grothendieck, Théorèmes de dualité pour les modules cohérents, Séminaire Bourbaki, No. **149**, Secr. Math. I.H.P., Paris, 1957.

[4] A. Grothendieck, Cohomologie locale des faisceaux cohérents et Théorèmes de Lefschetz locaux et globaux, Exp. XI, SGA 2, North Holland, Amsterdam, 1968.

[5] R. Hartshorne and A. Ogus, On the factoriality of local rings of small embedding dimension, Comm. Algebra **1** (1974), 415–437.

[6] J. Herzog, A. Simis and W. V. Vasconcelos, On the arithmetic and homology of algebras of linear type, Trans. Amer. Math. Soc. **283** (1984), 661–683.

[7] J. Herzog, A. Simis and W. V. Vasconcelos, Arithmetic of normal Rees algebras, J. Algebra **143** (1991), 269–294.

[8] M. Hochster, Properties of noetherian rings stable under general grade reduction, Arch. Math. **24** (1973), 393–396.

[9] C. Huneke, Strongly Cohen–Macaulay schemes and residual intersections, Trans. Amer. Math. Soc. **277** (1983), 739–763.

[10] C. Huneke and M. E. Rossi, The dimension and components of symmetric algebras, J. Algebra **98** (1986), 200–210.

[11] P. Samuel, Anneaux gradués factoriels et modules réflexifs, Bull. Soc. Math. France **92** (1964), 237–249.

[12] A. Simis and N. V. Trung, Divisor class group of ordinary and symbolic blow–ups, Math. Z. **198** (1988), 479–491.

[13] A. Simis, B. Ulrich and W. V. Vasconcelos, Jacobian dual fibrations, American J. Math. **115** (1993), 47–75.

[14] A. Simis and W. V. Vasconcelos, On the dimension and integrality of symmetric algebras, Math. Z. **177** (1981), 341–358.

[15] W. V. Vasconcelos, The complete intersection locus of certain ideals, J. Pure and Applied Algebra **38** (1985), 367–378.

[16] W. V. Vasconcelos, Modules of differentials of symmetric algebras, Arch. Math. **56** (1990), 436–442.

[17] J. Weyman, Resolutions of the exterior and symmetric powers of a module, J. Algebra **58** (1979), 333–341.

Splitting Properties of Extensions of the Wedderburn Principal Theorem

JOSEPH A. WEHLEN   Computer Sciences Corporation,
Integrated Systems Division, P. O. Box 1038, 304 West Route 38,
Moorestown, NJ 08057-0902
e-mail address: jwehlen@isd.csc.com

INTRODUCTION:

A traditional method for studying the structure of module-finite
associative algebras has been to separate the algebra into a
separable subalgebra and a radical subalgebra. When the base ring
R is a field and A is a module-finite R-algebra, the classical
result is the Wedderburn Principal Theorem: if A modulo its
Jacobson radical $J(A)$ is separable, then A contains a separable
subalgebra S such that $A = S + J(A)$ and $S \cap J(A) = (0)$. The
Mal'cev Conjugacy Theorem asserts that if S and S' are both
separable subalgebras of A such that $A = S + J(A) = S' + J(A)$,
then there is an x in $J(A)$ such that $S' = (1+x)S(1+x)^{-1}$.
   The purpose of this paper is to investigate conditions under
which both the summand and the intersection properties of the
Wedderburn Principal Theorem can be extended to more general base
rings with the prime radical substituted for the Jacobson
radical.
   The main result of Section 2 characterizes those rings R
with prime radical $L(R)$ for which every module-finite R-algebra A
contains a separable subalgebra S such that $A = S + L(A)$ as
$R/L(R)$-modules and $S \cap L(A) = (0)$ as those commutative π-regular
rings for which the natural map $\rho: R \to R/L(R)$ splits as $R/L(R)$-
modules. In Section 3, we investigate various conditions for

223

Noetherian (and almost Noetherian) ground rings under which the
trivial intersection between the separable subalgebra and the
prime radical holds. In Section 4, we extend results of W. C.
Brown (1971) and those of Section 3 to the non-Noetherian case.

Conventions: R denotes a commutative ring. All rings have an
identity and subrings contain the identity of the over ring. All
modules and all ring homomorphisms are unitary. By a module-
finite (projective) algebra A over R, we mean that A is finitely
generated (projective) as a module over R. J(-) denotes the
Jacobson radical of the ring (-) while L(-) denotes its prime
radical (Baer lower radical). To say that a commutative ring is
connected means that R has no idempotents but 0 and 1 [**cf.**
Demeyer & Ingraham, 1970, Proposition 4.7, p. 28]. By a semi-
local ring R we mean that R contains only a finite number of
maximal ideals (no chain conditions are implied).

   X(R) denotes the Pierce decomposition space of the
commutative ring R [see (Magid, 1974) or (Pierce, 1967) for
details]. Following the naming conventions of Magid (1974), we
say that a commutative ring R is **componentially a property** α
ring if R_x has property α for each x in **X**(R).

   A commutative ring R is said to be a **Hilbert ring** (or
Jacobson ring) if every prime ideal of R is an intersection of
maximal ideals of R.

   The Hochschild dimension of the R-algebra A, R-dim A, is the
projective dimension of A as a left $A^e = A\otimes_R A^{op}$-module. An R-
algebra A is said to be separable if R-dim A = 0. An algebra A
is an Azumaya algebra if it is separable as an algebra over its
center.

0. BACKGROUND

   E. C. Ingraham (1966) defined a commutative ring to be an
inertial coefficient ring if every module-finite R-algebra A,
with A modulo its Jacobson radical separable over R, contains a
separable subalgebra S such that A = S + J(A) as R-modules [the
sum need not be direct]. A separable subalgebra S with the above
property is called an **inertial subalgebra.** We shall say that the
trivial intersection property holds if the inertial subalgebra
intersected with the radical is zero. The Wedderburn Principal
Theorem, described anachronistically, states that a field is an
inertial coefficient ring with the trivial intersection property.
The development of the Hochschild cohomology (Hochschild, 1945)
spurred renewed interest in this theorem.

   Goro Azumaya (1951) provided a major breakthrough in
studying the structure of associative algebras over general
commutative rings using this paradigm. He proved that every

local Hensel ring is an inertial coefficient ring. [He provided
an alternate proof of this result in (Azumaya, 1971).] An
indication of the importance of the Henselian property for
inertial coefficient rings was provided by W. C. Brown and
Ingraham (1970) who showed that a semi-local ring R is an
inertial coefficient ring if and only if R is a finite direct sum
of local Hensel rings.

Additional early work along these lines was done by Charles
Curtis (1954), Chester Feldman (1951), and Daniel Zelinsky (1954)
in case the algebra is complete in the topology specified by its
Jacobson radical.

Using Azumaya's theorem in conjunction with the technique of
lifting separability idempotents from an algebra modulo its
radical to the algebra itself, Ingraham (1974) proved that every
Noetherian Hilbert ring is an inertial coefficient ring. Wehlen
(1990) showed that every commutative Hilbert ring is an inertial
coefficient ring.

Ingraham conjectured that a commutative ring R is an
inertial coefficient ring if and only if every module-finite R-
algebra A with Jacobson radical J(A) has the idempotent lifting
property, **viz.**, that every idempotent of A/J(A) is the image of
an idempotent of A. Ellen Kirkman (1976) proved the "only if"
portion of this conjecture, showing an inertial coefficient ring
is a Hensel ring in the sense of Greco (1968). The "if" portion
of Ingraham's conjecture remains open.

Since the prime radical (Baer lower radical) is a nil ideal,
Wehlen (1973) defined a commutative ring R to be a **weak inertial
coefficient ring** (abbreviated WIC-ring) if every module-finite R-
algebra A, with A modulo its prime radical L(A) separable over R,
contains a separable subalgebra S such that A = S + L(A) as R-
modules. He showed in (Wehlen, 1990) that every commutative ring
is a WIC-ring.

Ingraham (1974, Corollary 3.1) showed that the Mal'cev
Conjugacy Theorem holds for every commutative ring R relative to
the prime radical.

In all of the above work, the concern was with showing the
existence and conjugacy of inertial subalgebras. The trivial
intersection property was ignored. The traditional paradigm is
most completely satisfied when the trivial intersection property
holds. Brown (1970, 1972) characterized those rings for which
every module-finite R-algebra A with A modulo its Jacobson
radical separable over R contains an R/J(R)-separable subalgebra
S such that A = S + J(A) and S ∩ J(A) = (0) where the summation
is as R/J(R)-modules. Brown (1971) provided additional cases in
which the trivial intersection property holds.

1. DEFINITIONS OF THE SPLITTING PROPERTIES

The trivial intersection property of the Wedderburn
Principal Theorem implies that the sequence

$$0 \to J(A) \to A \to A/J(A) \to 0$$

splits as a sequence of modules over the base ring. Brown refers
to these extensions of the Wedderburn Principal Theorem as
"satisfying a splitting property." The splitting property of the
Wedderburn Principal Theorem relative to the prime radical may be
studied from several different viewpoints. Conditions may be
imposed upon the base ring, upon the separable homomorphic image,
or upon the algebra itself. Ingraham [1966, Lemma 2.3] has shown
that the following generic splitting property holds from the
viewpoint of the separable homomorphic image.

Result 1.1: Let A be a module-finite algebra over the
commutative ring R. If A/I is a projective, separable R-algebra
with $I \subseteq J(A)$ and A contains an R-separable subalgebra S such
that A = S + I, then $S \cap I = (0)$.

Combining this result with the fact that every commutative
ring is a WIC-ring [Wehlen, 1990, Corollary 1.4.1], we have the
following splitting property:

Corollary 1.1.1: Let A be a module-finite R-algebra such
that A modulo its prime radical L(A) is a projective, separable
R-algebra. A contains a separable subalgebra S such that A = S +
L(A) and $S \cap L(A) = (0)$.

This approach yields little information relative to the base
ring. W. C. Brown (1970) defined a splitting property for
commutative rings upon which he based his splitting conditions
for inertial coefficient rings.

Definition 1.2: A commutative ring R is said to be *split
relative to the ideal I* if there exists a homomorphism **j**: R/I →
R which splits the natural map ρ: R → R/I. The map **j** induces an
R/I-structure on R and hence on any R-algebra A.

Using Definition 1.2, Brown defined the following two
splitting properties. (The homomorphism **j** in Definition 1.2 is
assumed and will not be explicitly stated.) We shall say that an
R-algebra A is *quasi-R-connected* if every idempotent of A is
contained in R.

Definition 1.3: Let R be a commutative ring split relative
to its Jacobson radical J(R).
   (a) R is said to be a *strong IC-ring* (strong inertial
coefficient ring) if every module-finite R-algebra A with A/J(A)
separable over R contains a separable R/J(R)-subalgebra S such
that A = S + J(A) as R/J(R)-modules and $S \cap J(A) = (0)$. [Brown,
1970].

(b) R is said to be an *f-split IC-ring* if every module-finite, faithful, quasi-R-connected R-algebra A with A/J(A) R-separable contains a separable R/J(R)-subalgebra such that A = S + J(A) as R/J(R)-modules and S $\cap$ J(A) = (0). [Brown, 1971].

Brown (1972) completely characterized the strong IC-rings as those commutative rings each of whose stalks is a local Hensel ring split relative to its Jacobson radical. Brown (1971) showed that every Noetherian, integrally closed, Hilbert domain is an f-split IC-ring. (Brown used the terminology "the splitting property" for a connected f-split IC-ring.) Moreover, he proved that a local ring R split relative to its Jacobson radical is an f-split IC-ring if and only if R is a strong IC-ring [Brown, 1972, Corollary to Theorem 2].

In this paper we will examine several splitting properties for WIC-rings. The first two properties defined below are the WIC analogs of Brown's properties.

Definition 1.4: Let R be a commutative ring split relative to its prime radical.

(a) R is said to be a **split WIC-ring** if every module-finite R-algebra A with A/L(A) separable over R contains a separable R/L(R)-subalgebra S such that A = S + L(A) as R/L(R)-modules and S $\cap$ L(A) = (0).

(b) R is said to be an **f-split WIC-ring** if every module-finite, faithful, quasi-R-connected R-algebra A with A/L(A) separable over R contains a separable R/L(R)-subalgebra S such that A = S + L(A) as R/L(R)-modules and S $\cap$ L(A) = (0).

It is easy to verify that the class of split WIC-rings is closed under the "operations" of finite direct sums, homomorphic images, and module-finite commutative separable extensions [**cf.** Brown, 1971, Propositions 1-5].

A modification of Theorem 1.2 of Kirkman (1978) reduces the determination of all split WIC-rings to the case of a base ring R with no proper idempotents. Note first that a ring R is split relative to its prime radical by a homomorphism **j** if and only if R_x is split relative to its prime radical by j_x for each x in its Pierce decomposition space **X**(R).

Proposition 1.5: Let R be a commutative ring split relative to its prime radical L(R). A module-finite R-algebra A with A/L(A) separable over R contains an R/L(R)-subalgebra S such that A = S + L(A) and S $\cap$ L(A) = (0) if A_x contains a module-finite $R_x/L(R_x)$-subalgebra S'(x) such that A_x = S'(x) + L(A_x) and S'(x) $\cap$ L(A_x) = (0) for every x in **X**(R).

Proof: Let A be a module-finite R-algebra with A/L(A) separable over R; let R be split relative to its prime radical L(R) by **j**. Let $a_1, \ldots, a_n$ be the generators of A over R. Note that since R_x is a flat R-module (Magid, 1974, Proposition II.18), $A_x/L(A)_x \approx [A/L(A)]_x$ and $L(A)_x \subseteq L(A_x)$. By assumption, A_x contains a module-finite $R_x/L(R_x)$-separable subalgebra $S'(x)$ such that $A_x = S'(x) + L(A_x)$ and $S'(x) \cap L(A_x) = (0)$ for each x in $\mathbf{X}(R)$. There are elements $s_1(x),\ldots,s_{n(x)}(x)$ such that $s_1(x)_x,\ldots,s_{n(x)}(x)_x$ generate $S'(x)$. Let $S(x)$ be the $R/L(R)$-submodule generated by the $\{s_i(x)\}$. Then $S(x)_x = S'(x)$.

Since $S(x)_x$ is a module-finite $R_x/L(R_x)$ with identity 1_x, there are elements $r_{ijk}(x)$, $r_i(x)$ in $R/L(R)$ such that for $i,j = 1,\ldots,n(x)$:

(1) $(s_i(x))_x(s_j(x))_x = \sum_{k=1}^{n(x)} (r_{ijk}(x))_x(s_k(x))_x;$

(2) $1_x = \sum_{i=1}^{n(x)} (r_i(x))_x(s_i(x))_x.$

Since $A_x = S'(x)_x + L(A_x)$, it follows that there are elements $z_q(x)$ in $L(A)$ and elements $t_{qj}(x)$ in $R/L(R)$ such that for $q = 1,\ldots,n$:

(3) $(a_q)x = (z_q(x))_x + \sum_{j=1}^{n(x)} (t_{qj}(x))_x(s_j(x))_x.$

Finally, since $S(x)_x$ is $R_x/L(R_x)$-separable, there exist by (Demeyer & Ingraham, Proposition 1.1 (iii), p. 40) and (Kirkman, 1978, Lemma 1.1) elements $v_h(x)$, $w_h(x)$ in $S(x)$ for $h = 1,\ldots,m(x)$ and $r_{hj}(x)$, $r'_{hj}(x)$ in R such that:

(4) $(v_h(x))_x = \sum_{j=1}^{n(x)} (r_{hj}(x))_x(s_j(x))_x;$

(5) $(w_h(x))_x = \sum_{j=1}^{n(x)} (r'_{hj}(x))_x(s_j(x))_x;$

(6) $1_x = \sum_{h=1}^{m(x)} (v_h(x))_x(w_h(x))_x;$

(7) $\sum_{h=1}^{m(x)} [(s_j(x))_x(v_h(x))_x \otimes_{R_x/L(R_x)} (w_h(x))_x]$
     $= \sum_{h=1}^{m(x)} [(v_h(x))_x \otimes_{R_x/L(R_x)} (w_h(x))_x(s_j(x))_x]$
     in $S(x)_x \otimes_{R_x/L(R_x)} S(x)_x^{op}$ for all j.

Since these equations hold at x, they hold on some neighborhood U_x of x. Since $X(R)$ is compact, we may find a finite covering of $X(R)$ consisting of disjoint closed and open sets $\{N_i\}$ $i = 1,\ldots,n'$. Using the standard techniques of Magid (1974), we may piece together global elements

$r_{hj}, r'_{hj}, r_{ijk}, r_i, t_{qj} \in R/L(R)$; $s_i, v_h, w_h \in S$; $z_q \in L(A)$ satisfying equations (1) through (7) above for every x in $X(R)$. The subalgebra S generated by the $\{s_i\}$ is a module-finite, $R/L(R)$-separable subalgebra of A (see also, Kirkman & Kuzmanovich, 1978, Proposition 1.1). To see that S satisfies $S + L(A) = A$ as $R/L(R)$-modules, we have the following equations:

(8) $a = \sum_{i=1}^n v_i a_i$ with v_i in R since $\{a_i\}$ generates A over R;

     $= \sum_{i=1}^n (v'_i + x_i) a_i$ where $v'_i = \mathbf{j}(\rho(v_i))$ and x_i is in $L(R)$;

     $= \sum_{i=1}^n v'_i a_i + \sum_{i=1}^n x_i a_i$

     $= \sum_{i=1}^n v'_i [z_i + \sum_{j=1}^n r_{ij} s_j] + \sum_{i=1}^n x_i a_i$ by (3)

     $= \sum_{i=1}^n \sum_{j=1}^n v'_i r_{ij} s_j + [z_i + \sum_{i=1}^n x_i a_i]$ is in $S + L(A)$.

To see that $S \cap L(A) = (0)$, note that, since R_x is a flat R-module,

$(S \cap L(A))_y \subseteq S_y \cap L(A)_y \subseteq S_y \cap L(A_y) = (0)$

for every $y \in X(R)$. Hence $S \cap L(A) = (0)$. ∎

An R-module M is said to be **componentially projective** if M_x is R_x-projective for every x in **X**(R). A componentially projective R-module is R-flat.

Corollary 1.5.1: Let A be a module-finite R-algebra such that A modulo its prime radical L(A) is a separable, componentially projective R-algebra. Then A contains an R-separable subalgebra such that A = S + L(A) and S ∩ L(A) = (0).
Proof: This follows from 1.5 and 1.1.1.

A. Magid [1974, following Definition IV.1, pp. 79-80] provided the following example of a module-finite, commutative, separable, componentially projective R-algebra S which is not R-projective. Let X be the convergent sequence {1,½,..,1/n,...,0}. Let S be the set of continuous functions C(X,$\mathbb{C}$) from X to the complex numbers $\mathbb{C}$ with the discrete topology. Let R = {f∈S | f(0)∈$\mathbb{R}$}. Then $R_x = S_x = \mathbb{C}$ for x ≠ 0; $S_0 = \mathbb{C}$; while $R_0 = \mathbb{R}$. S is generated over R as an R-module by 1 and the constant function f(x) = i for all x in X. S is R-separable and componentially R-projective, but not R-projective.

Corollary 1.5.2: Let R be a commutative ring split relative to its Jacobson radical J(R). A module-finite R-algebra A with A/J(A) separable over R contains an R/L(R)-subalgebra S such that A = S + J(A) and S ∩ J(A) = (0) if A_x contains a module-finite $R_x/J(R_x)$-subalgebra S'(x) such that A_x = S'(x) + J(A_x) and S'(x) ∩ J(A_x) = (0) for every x in **X**(R).
Proof: Replace the prime radical with the Jacobson radical in the proof of 1.5.

We obtain as a corollary a complete characterization of the componentially semi-local f-split IC-rings as follows:

Corollary 1.5.3: Let R be a componentially semi-local ring. The following are equivalent:
    (a) R is an f-split IC-ring;
    (b) R is a strong IC-ring;
    (c) R_x is a split local Hensel ring for every x in **X**(R).
Proof: The equivalence of (b) and (c) follows from [Brown, 1972, Theorem 2]. To see the equivalence of (a) and (b), apply Corollary 1.5.2 in conjunction with [Brown, 1971, Corollary to Theorem 2], and Definition 1.3 (b).

2. SPLIT WIC-RINGS ARE π-REGULAR.

The purpose of this section is to characterize completely the split WIC rings. Recall that a ring R is said to be π - **regular** if for every a in R there is an x in R and a natural number n such that $a^n = a^n x a^n$. We will show that a commutative ring R split relative to its prime radical is a split WIC-ring if and only if R is a π-regular ring. We begin with two lemmas concerning π-regular rings, Hilbert rings, and Hensel rings.

Lemma 2.1: A local Hilbert ring is a Hensel ring.

Proof: By [Nagata, 1962, Theorem 43.15, page 185], R is a Hensel ring if and only if every module-finite extension T of R is a direct sum of local rings. Let R be a local Hilbert ring; then every module-finite extension T of R is a Hilbert ring and therefore L(T) = J(T). Thus there is a one-to-one correspondence between idempotents in T and T/J(T) since J(T) is nil. Hence T is a direct sum of local rings and so R is a Hensel ring. ∎

A ring R is said to be an **I-ring** if every non-nil left ideal contains a non-zero idempotent [Jacobson, 1964, p. 210]. In particular, if R is commutative, the prime radical and the Jacobson radical are equal [Burgess & Stephenson, 1979, p. 161]. A ring B is said to be **homomorphically an I-ring** (or **HI-ring**) if every homomorphic image of B, including B itself, is an I-ring. The following is an extension of a commutative version of a result of Burgess and Stephenson.

Lemma 2.2: Let R be a commutative ring. The following are equivalent:

    (i) R is an HI-ring.

    (ii) R is a π-regular ring.

    (iii) R_x is a local π-regular ring for each x in $\mathbf{X}(R)$.

    (iv) R_x is a local Hilbert ring for each x in $\mathbf{X}(R)$.

Proof: The equivalence of (i), (ii), and (iii) is demonstrated in [Burgess & Stephenson, 1979, Proposition 3.3]. We will show the equivalence of (iii) and (iv). In a commutative local ring S, every element is either contained in the maximal ideal or is a unit. Since any element t contained in the maximal ideal is nilpotent if S is a local Hilbert ring, there is an integer n such that $t^n = 0$ and so $t^n = t^n 1 t^n$. If t is a unit, then $t = t t^{-1} t$. Hence S is π-regular. Conversely, if S is a local π-regular ring, then S is an HI-ring and so by [Burgess & Stephenson, 1979, p. 163] J(S) = L(S) and so every prime ideal is an intersection of maximal ideals. Hence R is a local Hilbert ring. ∎

Proposition 2.3: Let R be a commutative ring split relative to its prime radical.

(a) R is a split WIC-ring if and only if R_x is a split WIC-ring for every x in $\mathbf{X}(R)$.

(b) A semi-prime ring R is a split WIC-ring if and only if R is von Neumann regular (absolutely flat).

(c) A semi-local, connected, commutative ring R is a split WIC-ring if and only if R is a local Hilbert ring.

Proof: (a) is an easy consequence of 1.5.

To prove the "only if" portion of (b) and (c), assume first that R is a split semi-prime WIC-ring. Let a be a non-unit of R and set I = aR. Setting $A = R/I^2$, it is clear that A is a module-finite R-algebra with prime radical L(A) containing I^2. A/L(A), being a homomorphic image of R, is R-separable. Since R is a split WIC-ring, the exact sequence

$$0 \to L(A) \to A \to A/L(A) \to 0$$

splits as an R-module. If j^* denotes the splitting map, then $j^*(1) = 1$ and, since R is semi-prime, $j^*(A/L(A)) = A$. Thus L(A) = (0) and so $I = I^2$. Consequently, there is a y in I such that a $= a^2y$. Since "a" was an arbitrary non-unit, R is a von Neumann regular ring. This completes the "only if" portion of (b).

To complete the "only if" portion of (c), note that when R is a semi-prime, connected, von Neumann regular ring, R is a field. Hence, for an arbitrary semi-local, connected, split WIC-ring R, R/L(R) must be a field. So R must be a local ring with J(R) = L(R). Since R is local with nil maximal ideal, every prime ideal is maximal and so R is a Hilbert ring.

For the "if" portion of (b), if R is von Neumann regular, then Proposition 1.5 guarantees that R is a split WIC-ring since R_x is a field which is a split WIC-ring by the Wedderburn Principal Theorem.

For the "if" portion of (c), let R be a local Hilbert ring, split relative to its prime radical. By Lemma 2.1, R is a Hensel ring; since R is split, R is a strong IC-ring (Brown, 1970, Theorem 1). Now if A is a module-finite R-algebra with A/L(A) separable over R, A/L(A) is a separable algebra over the field R/L(R) and hence is semi-simple. Thus J(A) = L(A) and so R is a split WIC-ring. ∎

Combining the results of the above proposition, we have the characterization of split WIC-rings in terms of the Pierce stalks of R.

Corollary 2.3.1: A commutative ring R is a split WIC-ring if and only if every stalk is a local Hilbert ring which is split relative to its prime radical.

Combining 2.3.1 and 2.2, we have the proposed characterization of split WIC-rings.

Theorem 2.4: Let R be a commutative ring split relative to its prime radical. R is a split WIC-ring if and only if R is a π-regular ring.

A commutative ring R is said to be **locally perfect** if R is von Neumann regular modulo a T-nilpotent ideal [Burkholder, 1986, Remark 3.5(a)]. A commutative ring R is locally perfect if and only if R is componentially perfect [Burkholder, 1989, Proposition 7] if and only if R is a P-exchange ring [Kambara & Oshiro, 1988, Corollary 1 to Theorem 2]. It is easy to see that a local perfect ring is a local Hilbert ring. Thus

Corollary 2.4.1: A commutative P-exchange ring R split relative to its prime radical is a split WIC-ring.

Now every Henselian domain R split relative to its Jacobson radical (such as R = F[[X]] where F is a field) is a strong IC-ring (Brown, 1970). However, by the argument of 2.3(b) R is not a split WIC-ring even though R is split relative to its prime radical by the identity map. Thus the class of split WIC-rings is a proper subclass of the strong IC-rings.

3. f-SPLIT IC-RINGS & f-SPLIT WIC-RINGS: THE NOETHERIAN CASE.

In this section, we give several classes of Noetherian and "almost" Noetherian rings which are f-split WIC-rings. We note first that any domain is split relative to its prime radical by the identity map. This fact will be used throughout without further comment.

Brown (1971, Theorem 3) showed that a Noetherian integrally closed Hilbert domain is an f-split IC-ring. He also showed (1971, Example 3.2) that R = Q[x]$_{(x)}$[Y] is a Noetherian integrally closed domain which is not a Hilbert ring and gave an example of a module-finite R-algebra to show that R is not an f-split IC-ring. However, R is an f-split WIC ring as proved below:

PROPOSITION 3.1: A Noetherian integrally closed domain is an f-split WIC-ring.
Proof: Let A be a faithful, module-finite algebra with no proper idempotents over the Noetherian integrally closed domain R with A/L(A) R-separable. The result will be proved if we can show that A/L(A) is a R-projective R-algebra for then Corollary 1.1.1 guarantees that the weak inertial subalgebra exists and has the desired properties.

Since the prime radical is a nil ideal, $L(A) \cap R \subseteq L(R) = (0)$ and so $A/L(A)$ is a faithful R-algebra. Since A has no proper idempotents and idempotents lift modulo the nil ideal $L(A)$, $A/L(A)$ and its center $C = \text{Cen}(A/L(A))$ are faithful, module-finite R-algebras with no proper idempotents and hence must be torsion-free. By [Janusz, 1966, Corollary 4.2], C is projective over R; so $A/L(A)$ being an Azumaya C-algebra is projective over C and hence over R.

Hence R is an f-split WIC-ring. ∎

For a commutative Noetherian Hilbert ring R, $L(A) = J(A)$ for every module-finite R-algebra A [Ingraham, 1969, Corollary 1]. Thus we have as a corollary to the proof of 3.1:

COROLLARY 3.1.1: [Brown, 1971, Theorem 3] A Noetherian, integrally closed, Hilbert domain is an f-split IC-ring.

A commutative ring R is said to be *locally Noetherian* if each localization of R at a prime ideal is a Noetherian ring. An almost Dedekind domain is an example of an integrally closed, locally Noetherian domain.

COROLLARY 3.1.2: A locally Noetherian, integrally closed domain is an f-split WIC-ring.

Proof: In order for the proof of 3.1 to hold for R locally Noetherian, we must show that the center of a torsion-free separable R-algebra with no proper idempotents is R-projective to replace the reference to [Janusz, 1966, Corollary 4.2].

Let R be an integrally closed, locally Noetherian domain and S a commutative torsion-free separable R-algebra with no proper idempotents. Let $\wp$ be a prime ideal of R. Then $S_\wp$ is a torsion-free, separable algebra over the Noetherian integrally closed domain $R_\wp$. By [Janusz, 1966, Theorem 4.3] $S_\wp$ is $R_\wp$-projective. Since this is true at every prime ideal $\wp$, S is a torsion-free, flat, module-finite R-module and so is R-projective by [Cartier, 1958, Appendix, Lemma 5; Endo, 1962, Theorem 2]. ∎

A commutative, not necessarily Noetherian domain is said to be a *Prüfer domain* if every finitely generated, torsion-free R-module is projective. A Prüfer domain need not be even locally Noetherian, but a Prüfer domain is integrally closed.

COROLLARY 3.1.3: A Prüfer domain is an f-split WIC-ring.

Proof: Replace the reference to [Janusz, 1966, Corollary 4.2] in the proof of 3.1 with a reference to [Wehlen, 1975, Theorem 1.1]. ∎

Before we proceed to the next result, we note that for a locally Noetherian, integrally closed domain R and a commutative,

module-finite, torsion-free, separable R-algebra S with no proper
idempotents, then S is R-projective from the proof of 3.1.2.
From Proposition II.2.3 & Corollary III.2.6 of Demeyer and
Ingraham (1970), S is a domain. It is easy to verify that S is
locally Noetherian and integrally closed.

For the above integrally closed domains, the trivial
intersection property extends to algebras with proper idempotents
if we assume that the algebra is torsion-free.

PROPOSITION 3.2: Let R be a commutative integrally closed
domain which satisfies one of the following conditions: (a) R is
Noetherian; (b) R is locally Noetherian; or (c) R is Prüfer. If
A is any module-finite R-algebra with A/L(A) R-separable and
torsion-free over R, then A contains a separable subalgebra S
such that A = S + L(A) and S $\cap$ L(A) = (0).

Proof: Let R be a ring of one of the above types and let A
be a module-finite R-algebra such that A/L(A) is torsion-free and
separable over R. Then C = Cen(A/L(A)) is a direct sum of R-
projective domains. {For case (a), see the proof of [Wehlen,
1973, Theorem 2.6]; for case (b), see the note proceeding this
proposition; for case (c), see Theorem 2.4 in (Wehlen, 1975).}
Now A/L(A) is an Azumaya C-algebra and so is C-projective. This
ensures that A/L(A) is R-projective and thus by Corollary 1.1.1
the result holds. ■

Let R = $Z_S[5^{1/2}]$ be the ring of rational integers Z
localized at the set
$$S = \{ p^i \mid i = 0,1, \ldots; p \neq 2; p \neq 5; p \text{ a prime}\}$$
and then with the square root of 5 adjoined. Brown (1971) proved
that this ring is a Noetherian Hilbert domain which is not
integrally closed and that the R-algebra A = $Z_S[5^{1/2}][X]/(x^2+x-1)$
is a module-finite R-algebra. Moreover, B = $A/(1+5^{1/2}+2x)^2$ is a
module-finite R-algebra which has radical
$$J(B) = L(B) = (1+5^{1/2}+2x)/(1+5^{1/2}+2x)^2$$
and B/J(B) is R-separable; however, B has no separable
subalgebra adding with the prime radical to yield B; hence R is
neither an f-split IC-ring nor an f-split WIC-ring. This example
shows that integral closure (or some similar condition) is also
required for a domain to be an f-split WIC-ring.

The results of Proposition 3.1 can be extended to rings with
non-zero prime radical as in the following proposition. The
proof of this result follows as in [Brown, 1971, Theorem 4] since
the prime radical is nilpotent for a Noetherian ring.

A commutative ring is said to be a *p.p. ring* if every
principal ideal is projective. A p.p. ring is said to be
integrally closed if it is integrally closed in its total
quotient ring. Details concerning commutative p.p. rings may be
found in (Bergman, 1971).

PROPOSITION 3.3: Let R be a Noetherian ring split relative to its prime radical L(R). If R/L(R) is an integrally closed p.p. ring, then R is an f-split WIC-ring.

Proof: By 1.5, we may assume that R is a connected ring with prime radical L(R). Since we are concerned with a quasi-R-connected algebra A, we may let A be a module-finite R-algebra with no proper idempotents such that A/L(A) is separable. For any commutative ring R, A = S + L(A) for some R-separable subalgebra S of A. By the hypotheses on A, S also has no proper idempotents. One can readily verify that S ∩ L(A) = L(S); and so it suffices to show that S can be further decomposed as an R/L(R)-separable subalgebra S' and that S = S' + L(S) with S' ∩ L(S) = (0).

To prove this result, we may consider the sequence of R/L(R)-modules:

$$0 \to L(S) \to S \to S/L(S) \to 0.$$

Let us assume first that $L(S)^2 = (0)$. Then there exists an R/L(R)-separable subalgebra S˜ of S such that S = S˜ + L(S) and S˜ ∩ L(S) = (0) [see, e.g., Proposition 12 of Azumaya (1972)].

Let us proceed by induction. Suppose that whenever the $L(S)^m = (0)$ for $2 \le m < n$, there exists an R/L(R)-separable subalgebra S˜ of S such that S˜ + L(S) = (0) and S˜ ∩ L(S) = (0). Consider the exact sequence

$$0 \to L(S)/L(S)^2 \to S/L(S)^2 \to S/L(S) \to 0.$$

By the inductive hypothesis there is an R/L(R)-separable subalgebra $T \subset S/L(S)^2$ such that $S/L(S)^2 = T + L(S)/L(S)^2$ and $T \cap L(S)/L(S)^2 = (0)$. Since T is isomorphic to S/L(S), consider the sequence:

$$S \to S/L(S) \approx T \to (0).$$

By the correspondence theorem, there is a subalgebra S' of S such that $L(S)^2 \subset S'$ and $S'/L(S)^2 = T$. Thus $L(S') = L(S)^2$. Now the sequence

$$0 \to L(S)^2 \to S' \to S/L(S) \to (0)$$

is exact and, by the inductive hypothesis, S' contains an inertial R/L(R)-subalgebra S˜ with the trivial intersection property $S˜ \cap L(S)^2 = (0)$. Hence S˜ ∩ L(S) = (0). S˜ is the desired inertial subalgebra of S. ■

4. f-SPLIT IC-RINGS & f-SPLIT WIC-RINGS: THE NON-NOETHERIAN CASE

The two examples of Brown (1971) - discussed in the previous section - show that, for an integral domain to be an f-split IC-ring, the conditions of integral closure and Hilbert domain are in some sense required. However, the Noetherian condition is not required as we will show in the following theorem.

In addition, the Noetherian property can be removed from Proposition 3.1. Two lemmas relating the Jacobson radical of an R-algebra A to the Jacobson radical of some subrings.

LEMMA 4.1: [Ingraham, private communication] Let A be an R-algebra and let R' be a subring of R. Suppose that B is a subring of A such that B is an R'-algebra and A is a finitely generated left B-module, and RB = A. Then $J(B) \subseteq J(A) \cap B$.

Proof: Suppose that $J(B) \not\subseteq J(A)$. Then there is a maximal left ideal $\mathcal{M}$ of A with $J(B) \not\subseteq \mathcal{M}$. Therefore $AJ(B) + \mathcal{M} = A$. Since R is in the center of A, $AJ(B) = RBJ(B) = J(B)BR = J(B)A$. It follows that $J(B)A + \mathcal{M} = A$. But $A/\mathcal{M}$ is finitely generated as a left B-module with $J(B)(A/\mathcal{M}) = A/\mathcal{M}$. By Nakayama's Lemma, $A/\mathcal{M} = (0)$ which is a contradiction. ∎

LEMMA 4.2: Let $S \subseteq R$ be commutative Hilbert rings. Let A be a finitely generated R-algebra defined by the following relations:
$$A = \sum_{i=1}^{n} Ra_i; \quad a_i a_j = \sum_{k=1}^{n} r_{ijk} a_k; \quad 1 = \sum_{i=1}^{n} t_i a_i.$$
If the subring S contains the coefficients r_{ijk}, t_i, and $B = \sum_{i=1}^{n} Sa_i$, then $J(B) = J(A) \cap B$.

Proof: Let x be an element of $J(A) \cap B$. Then AxA is a nil ideal of A; thus BxB is a nil ideal of B and so is contained in the Jacobson radical of B. The other inclusion follows from 4.1 since the construction of B guarantees that RB = A. ∎

THEOREM 4.3: Let R be an integrally closed integral domain (not necessarily Noetherian).
   (a) If R is a Hilbert domain, then R is an f-split IC-ring.
   (b) R is an f-split WIC-ring.
Proof: To prove (a), let A be a module-finite, faithful R-algebra with no proper idempotents. In particular, if $A = \sum_{i=1}^{n} Ra_i$, then
   (1) $a_i a_j = \sum_{k=1}^{n} r_{ijk} a_k$;
   (2) $1 = \sum_{i=1}^{n} t_i a_i$.
Then $A/J(A) = \sum_{i=1}^{n} R\psi(a_i)$ where ψ denotes the natural map from A to $A/J(A)$. Let P denote the prime ring of R and let T' denote the ring generated by P and the elements r_{ijk}, t_i. Then T' is a Noetherian Hilbert domain contained in R. Let T denote the integral closure of T'. T is then a Noetherian, integrally closed, Hilbert domain contained in R. [See Proposition 14.4 of (Nagata, 1962)]. Set $B^* = \sum_{i=1}^{n} T\psi(a_i)$. By Lemma 1.3 of (Wehlen, 1990), B^* is T-separable. Consider the algebra $B = \sum_{i=1}^{n} Ta_i$. Then B is a module-finite T-algebra since (1) and (2) hold in B. By Lemma 4.2, $B/J(B)$ is T-separable and may be

identified with a subring of A/J(A). Thus B* can be identified
by B/J(B) and so, by Lemma 1.3 of Wehlen (1990), B/J(B) is T-
separable. Moreover, B is a torsion-free, module-finite R-
algebra without proper idempotents. Since T is a Noetherian,
integrally closed, Hilbert domain, by Theorem 3 of Brown (1971),
B contains a module-finite, separable subalgebra S' such that B =
S' + J(A) and S' ∩ J(B) = (0). Set S = RS' and note that RJ(B)
⊆ J(A). S is separable since it is the homomorphic image of
R ⊗$_T$ S'. The following inclusions show that S is an inertial
subalgebra of A:

$$A = RB \subseteq R(S' + J(B)) \subseteq RS' + RJ(B) \subseteq S + J(A).$$

It remains to show that S ∩ J(A) = (0). Note that J(S) =
S ∩ J(A) since S is an inertial subalgebra [see Remark 2.3 of
(Ingraham, 1966)]. Since A is faithful over R and has no proper
idempotents, S is torsion-free over R. If S has a non-trivial
Jacobson radical J(S), J(S) is a nil ideal since R is a Hilbert
ring. However, since S is faithful and torsion-free, S is ·
faithfully contained in S ⊗$_R$ Q where Q is the field of quotients
of R. Since S ⊗$_R$ Q is a separable algebra over a field, it
contains no non-trivial nil ideals; hence S contains no non-
trivial nil ideals and J(S) = (0).

To prove (b), replace the Jacobson radical with the prime
radical in the above proof, use Lemma 1.2 of Wehlen (1990)
instead of 4.2 to see that L(B) = L(A) ∩ B for the algebra B in
the proof, and use Proposition 3.1 of this paper instead of
Theorem 3 of Brown (1971) for the existence of the separable
subalgebra S' of B. ■

COROLLARY 4.3.1: Let R be an integrally closed p.p. ring.
(a) If R is componentially a Hilbert domain, then R is an f-
split IC-ring.
(b) R is an f-split WIC-ring.
Proof: To prove (a), combine 1.5.2 with 4.3 (a). For (b),
combine 1.5 with 4.3(b).

COROLLARY 4.3.2: A commutative·semihereditary ring is an f-
split WIC-ring.
Proof: A commutative semihereditary ring is an integrally
closed p.p. ring which is componentially a Prüfer domain.

5. FINAL REMARKS AND A QUESTION

Alex Rosenberg and Daniel Zelinsky (1956) commented that
finding R-algebras A with Hochschild dimension less than or equal
to 1 (R-dim A ≦ 1) generalizes the Wedderburn Principal Theorem,
while finding algebras with R-dim A = 0 generalizes the Mal'cev
Conjugacy Theorem. The connection between the Hochschild

dimension and these theorems has been masked in this paper with the exception of the proof of 3.5. The connection is more readily apparent in Brown's work (1970, 1972).

In addition to the module-finite case studied in this paper, the Wedderburn-Mal'cev Theorems have been extended to the locally finite case by Rosenberg and Zelinsky (1956), R. Reisel (1956), and Ian Stewart (1976).

Recently, Goro Azumaya (1980) defined a ring A to be a **separable ring** [called an **Azumaya ring** by Burkholder (1986, 1989)] if (a) every homomorphic image of the center of A is the center of the homomorphic image and (b) every two-sided ideal of A is generated by central elements. Any simple ring is an Azumaya ring. Although there are Azumaya rings with arbitrary Hochschild dimension, can the Wedderburn Principal Theorem be extended to a ring B such that B modulo its Brown-McCoy radical G(B) is an Azumaya ring? Stated somewhat differently, under what restrictions on B, B/G(B), or the center of B/G(B) does there exist an Azumaya ring S such that $B = S + G(B)$ and $S \cap G(B) = (0)$?

Consider a ring $B = A \otimes_R D$ where A is a module-finite projective R-algebra with A/L(A) an Azumaya R-algebra and D an Azumaya ring containing a faithful copy of R in its center. Then there is a separable subalgebra S such that $A = S + L(A)$ and $S \cap L(A) = (0)$ by Result 1.1. It is easy to see that

$$B \approx A \otimes_R D \approx S \otimes_R D + L(A) \otimes_R D$$

as R-modules and hence as $R \otimes_R$ Cen(D)-modules. Thus B is a Cen(D)-module direct sum of an Azumaya ring and the Brown-McCoy radical of B.

REFERENCES:

1. G. Azumaya, On maximally central algebras, *Nagoya Math. J.* **2**(1951), 119-150.
2. G. Azumaya, Algebras with Hochschild dimension ≤ 1, *Ring Theory*, Academic Press (1971).
3. G. Azumaya, Separable rings, *J. Algebra*, **63**(1980), 1-14.
4. G. Bergman, Hereditary commutative rings and centres of hereditary rings, *Proc. Lond. Math. Soc.* **23**(1971), 214-271.
5. W. C. Brown, Strong inertial coefficient rings, *Mich. Math. J.* **17**(1970), 73-84.
6. W. C. Brown, Some splitting theorems for algebras over commutative rings, *Trans. Amer. Math. Soc.* **162**(1971), 103-113.
7. W. C. Brown, A splitting theorem for algebras over commutative von Neumann regular rings, *Proc. Amer. Math. Soc.* **36**(1972), 369-374.

8.   W. C. Brown and E. C. Ingraham, A characterization of semi-local inertial coefficient rings, *Proc. Amer. Math. Soc.* **26**(1970), 10-14.

9.   W. D. Burgess and W. Stephenson, Rings all of whose Pierce stalks are local, *Canad. Math. Bull.* **22**(1979), 159-164.

10. D. G. Burkholder, Azumaya rings with locally perfect centers, *J. Algebra* **103**(1986), 606-618.

11. D. G. Burkholder, Azumaya rings, Pierce stalks, and central ideal algebras, *Comm. Algebra* **17**(1989), 103-113.

12. P. Cartier, Questions de rationalite de diviseurs en geometrie algebrique, *Bull. Math. Soc. France* **86**(1958), 177-251.

13. C. W. Curtis, The structure of nonsemisimple algebras, *Duke Math. J.* **21**(1954), 79-85.

14. F. Demeyer and E. C. Ingraham, <u>Separable</u> <u>Algebras</u> <u>over</u> <u>Commutative</u> <u>Rings</u>, *Lecture Notes in Mathematics #181*, Springer-Verlag, Berlin-Heidelberg-New York (1970).

15. S. Endo, On flat modules over commutative rings, *J. Math. Soc. Japan* **14**(1962), 284-291.

16. C. Feldman, The Wedderburn Principal Theorem in Banach algebras, *Proc. Amer. Math. Soc.* **2**(1951), 771-777.

17. S. Greco, Algebras over nonlocal Hensel rings, *J. Algebra* **8**(1968), 48-56.

18. G. Hochschild, On the cohomology groups of an associative algebra, *Ann. Math.* **46**(1945), 568-579.

19. E. C. Ingraham, Inertial subalgebras of algebras over commutative rings, *Trans. Amer. Math. Soc.* **124**(1966), 77-93.

20. E. C. Ingraham, On the occasional equality of the lower and Jacobson radicals in Noetherian rings, *Arch. Math. (Basel)* **20**(1969), 267-269.

21. E. C. Ingraham, On the existence and conjugacy of inertial subalgebras, *J. Algebra* **31**(1974), 547-556.

22. N. Jacobson, "Structure of Rings," Amer. Math. Soc. Colloq. Publ., Vol. 37, Amer. Math. Soc., Providence, RI, 1964.

23. G. J. Janusz, Separable algebras over commutative rings, *Trans. Amer. Math. Soc.* **122**(1966), 461-479.

24. H. Kambara & K. Oshiro, On P-exchange rings, *Osaka J. Math* **25**(1988), 833-842.

25. E. E. Kirkman, Inertial coefficient rings & the idempotent lifting property, *Proc. Amer. Math. Soc.* **61**(1976), 217-222.

26. E. E. Kirkman, The Pierce representation of an inertial coefficient ring, *Rocky Mtn. Jnl. of Math.* **8**(1978), 533-538.

27. E. E. Kirkman and J. J. Kuzmanovich, Orders over hereditary rings, *J. Algebra* **55**(1978), 1-27.

28. A. R. Magid, *The Separable Galois Theory of Commutative Rings*, Marcel Dekker, 1974.

29. M. Nagata, *Local Rings*, Wiley-Interscience, New York, 1961.

30. R. S. Pierce, Modules over commutative regular rings, *Mem.
     Amer. Math. Soc. #70*, 1967.
31. R. B. Reisel, A generalization of the Wedderburn-Mal'cev
     theorem to infinite-dimensional algebras, *Proc. Amer. Math.
     Soc.* **7**(1956), 493-499.
32. I. Stewart, The Wedderburn-Mal'cev theorems in a locally
     finite setting, *Arch. Math* **27**(1976), 120-122.
33. J. A. Wehlen, Algebras over Dedekind domains, *Canad. J.
     Math.* **25**(1973), 842-855.
34. J. A. Wehlen, Algebras over absolutely flat commutative
     rings, *Trans. Amer. Math. Soc.* **196**(1974), 149-160.
35. J. A. Wehlen, Separable algebras over Prüfer domains, *Rocky
     Mtn. Jnl. of Math.* **5**(1975), 145-155.
36. J. A. Wehlen, A complete characterization of weak inertial
     coefficient rings, *Comm. in Algebra* **18**(1990), 3577-3588.
37. D. Zelinsky, Raising idempotents, *Duke Math. J.* **21**(1954),
     79-90.

Index

Printed and bound by CPI Group (UK) Ltd, Croydon, CR0 4YY
21/10/2024
01777093-0011